ABOUT THE AUTHOR

Dr. S.J. Singh received his M.Sc. (Agr.) and Ph.D. degree in the field of Plant Pathology from Banaras Hindu University. He joined Indian Council of Agricultural Research in 1966 and worked on potato viruses from 1966-1969 at C.P.R.I. Shimla. He joined Indian Institute of Horticultural Research, Hessaraghatta, Bangalore in 1969 and worked in several capacities including Head Division of Plant Pathology from 1969-1997. During his stay at Bangalore he worked on virus and phytoplasma diseases of horticultural crops. He is presently working as Principal Scientist and Head at I.A.R.I. Regional Station, Pune from 1997. Dr. Singh also served as Project Coordinator (Betelvine) from 1985-88. Dr. Singh was appointed as FAO National Consultant (Plant Protection) under the FAO—aided project on "Greenhouse Floriculture Technology for Small Scale Farmers" from 1998-2000. He is a Co-principal Investigator of a NATP funded project on "Diagnostics for virus and virus-like pathogens infecting citrus." He has published about 240 research articles in national and international journals. He is a well recognised scientist in the field of electron microscopy of viruses and phytoplasmas. He received training in operation and maintenance of Transmission Electron Microscope from JEOL, Tokyo, Japan and in "Advanced Methods in Virus Diagnosis" at AVRDC, Taiwan. Dr. Singh is the invited author of a book entitled "Betelvine cultivation in India" published by Ministry of Agriculture, Government of India and editor of the book on "Advances in diseases of fruit crops in India." He has also compiled two technical bulletins entitled "Viral and phytoplasma disease of vegetable crops" and "Diseases of horticultural crops and their management". He has contributed about 30 reviews/invited chapters in books. He has toured to several foreign countries namely—Japan, Taiwan, Thailand, Hongkong and Nepal. Dr. Singh is a recognised teacher and guide of post-graduate students in Plant Pathology, of the Mahatma Phule Krishi Vidyapeeth, Rahuri (Maharashtra). He was elected as President of Indian Phytopathological Society (Southern Zone) during 1989. Dr. Singh is a fellow of Indian Phytopathological Society (IPS) and life member of IPS, Indian Virological Society and Association of Plant Pathologists of India. Dr. Singh has been honoured by the American Biographical Institute with an honorary appointment to the Research Board of Advisors in the year 2001.

VIRAL DISEASES OF BANANA
—A HAND BOOK—

Dr. S.J. SINGH
M.Sc. (Ag.) Ph.D. (Pl. Path)., FPSI, ARS
Principal Scientist & Head
IARI Regional Station
Agricultural College Estate
Shivajinagar, Pune-411005

KALYANI PUBLISHERS
LUDHIANA • NEW DELHI • NOIDA (U.P.)
HYDERABAD • CHENNAI • KOLKATA • CUTTACK

KALYANI PUBLISHERS

Head Office

B-I/1292, Rajinder Nagar, Ludhiana-141 008

Administration Office

4779, Ansari Road,
23, Daryaganj, New Delhi-110 002

Works

B-16, Sector-8, NOIDA (U.P.)

Branches

1, Mahalakshmi Street,
T. Nagar, Chennai-600 017

110/111, Bharatia Towers, Badambadi,
Cuttack-9 (Orissa)

3-5-1108, Narayanaguda, Hyderabad-500 029

10/2B, Ramanath Mazumdar Street,
Kolkata-700 009

0094 03 04 06

First Edition, 2003

Typesetting at : **Times Printographic**
921, Jatwara Street, Tiraha Behram Khan
Daryaganj, New Delhi-110 002

ISBN 81-272-0817-5

PRINTED IN INDIA

Printed by: Xpress Grafics, Delhi-28
and Published by Mrs. Usha Raj Kumar for
Kalyani Publishers, New Delhi- 110 002

FOREWORD

Banana is one of the most important fruit crops of the world which is cultivated in approximately 120 countries mainly in the tropics and sub-tropics providing sustenance to millions of people. India has emerged as the largest producer of banana in the world with an estimated production of about 100 million tonnes. Banana production is being adversely affected due to various viral diseases namely, bunchy top, infectious chlorosis (Mosaic), streak virus and bract mosaic. Until mid 1980s, banana bunchy top and infectious chlorosis were recognized as a major threat to banana production, but later on two relatively recently identified viral diseases namely 'banana streak' and 'bract mosaic' have become prominent resulting in significant economic yield losses.

This hand book written by Dr. S.J. Singh is a source of comprehensive information devoted entirely on the various aspects of banana virus diseases. The information has been very well presented and supplemented with good photographs. This book would be of immense use to all the banana cultivators, teachers, students, extension workers and private companies.

I congratulate Dr. S.J. Singh

November 7, 2001

(PANJAB SINGH)
Secretary, Department of Agricultural Research & Education
& Director General, Indian Council of Agricultural Research
Krishi Bhawan, New Delhi

PREFACE

Banana is as old as Indian civilization and its antiquity can be traced from its mention in Ramayana (2009 B. C.); Kautilya's Arthashastra (300-400 B. C.) and its paintings and sculptures of Ajanta and Ellora. Megasthanes mentioned about ' Pala ' or 'Tala ' the fruit eaten by sages of ancient India, as seen by Alexander The Great. Banana has became a staple food due to its rich source of carbohydrate, vitamins and minerals. In India, banana has a great socio-economic significance. Banana is grown across the country in tropical and sub-tropical regions and is interwoven with the Indian culture and heritage.

No one will dispute the statement that banana viruses are responsible for considerable economic losses in productivity of banana. Losses are often more insidicius, frequently less conspicuous and therefore go unnoticed. Latent infections usually result in banana plant that grow more slowly, attain a smaller size, produce less fruit, alter product composition, reduce life span and act in various other ways that go unnoticed. The distribution of banana germplasm during the 20th century seems to be responsible for the long- distance spread of banana viruses.

Literature on banana viruses are scattered. Hence it was thought to bringout all the information at one platform and therefore there was a need for publication of a book exclusively on banana virus diseases.

This book contains all the relevant informations on banana virus diseases and has been classified in different head *viz.*, history and geographical distribution, symptomatology, transmission, etiology, physico-chemical and molecular properties of the viruses, epidemiology, host range, assessment of losses, bio-chemical changes and management. Attempts have been made to include up to date information on banana viruses and supplemented with good coloured photographs. This book would be very useful to virologists, teachers, students, farmers, private companies and extension workers.

I am very thankful to Dr. B. E. L. Lockhart, Department of Plant Pathology, University of Minnesota, U. S. A. for going through the manuscript, for his critical comments for the improvement of this book and for providing photographs of banana virus diseases. I am also thankful to Dr. R. Selvarajan, Senior Scientist, National Research Centre for Banana, Trichy for providing photographs of bract mosaic. My sincere thank are due to Dr. H. P. Singh, Horticulture Commissioner, Government of India, New Delhi for his encouragement, and suggestion for writing this book. Sincere thanks are also due to Dr. R. K. Singh and Dr. Satya Prakash, Senior Research Fellows at IARI Regional Station, Pune for their untiring help in the preparation of this book.

I am grateful to my wife Sumitra Singh and my children Deepak and Preeti for their continuous support and encouragement in bringing out this book.

A word of appreciation is due to M/s Kalyani Publishers, Ludhiana for doing an excellent job in publication of this book.

(S. J. SINGH)

CONTENTS

INTRODUCTION

Banana (*Musa* spp.) is one of the most important, but undervalued, food crops of the world and provides sustenance to millions of people. It is reported to be fourth most important global food commodity after rice, wheat and milk in terms of gross value of production. Banana is believed to be as old as cultivation itself and is among the oldest fruit cultivated by man. In India, this crop is inter-woven with national heritage, having great socio-economic significance and with multifarious uses as it is referred to "Kalpatru" (Plant of virtues).

The early chronology of the spread and cultivation is still principally a matter of conjecture. It is believed that banana originated in South East Asia. The earliest written record of banana cultivation are from India in the Epics of the Pali Buddhist canon of 500-600 BC (Reynolds, 1951). There are also written records of banana in Indonesia about 350 BC and China about AD 200.

Having been introduced to the various parts of the world, banana is Asia's most important contribution to the international fruit industry. It has become very important stable food particularly to the marginal segments of societies in much of the developing world. Banana is cultivated predominantly in gardens small holdings in some 120 countries, mainly in the tropics and sub-tropics. Roughly one-third of total world production of bananas, which in 1998 was estimated to be approximately 88 million tonnes (Anon, 1999) comes from the Asian Pacific region. Most fruit is consumed locally. It has been estimated that the highest consumption rates on the island of New Guinea and in the Great Lakes region of East Africa, where bananas occupy a large proportion of the diet and amount to 200-250 kg person^{-1} year^{-1}. This compares with consumption figures of approximately 15-16 kg person^{-1} year^{-1} in Europe and North America (Anon, 1992). Banana for export accounts for only about 10% of total production.

Taxonomically, banana belongs to the genus *Musa* of the family *Musaceae*, of the order *Zingiberales*. The genus *Musa*, which has about 40 species, is divided into five sections :

Eumusa and *Rhodochlamys*—contain species which have 22 (n = 11) chromosomes.

Australimusa and *Callimusa*—contain species which have 20 (n = 10) chromosomes.

Ingentimusa—contains one species which has 14 (n = 7) chromosomes.

Eumusa is the largest and most wide-ranging section of the genus and contains *Musa acuminata* Colla and *Musa balbisiana* Colla, which are the principal progenitors of most edible banana cultivars.

Banana can be divided into two main categories : Desert bananas and cooking bananas. Desert bananas, which constitute 43% of world production (Anon, 1992), are eaten raw when ripe. Among different types of desert bananas, fruit from the cultivars in the Cavendish subgroup are the most common. Cooking bananas, which account for the other 57% of world production (Anon, 1992), are usually starchy and ripe and need to be boiled, fried or roasted to make them palatable. Plantains, the most well known of the cooking type, form 23% of the world's total production of bananas (Anon, 1992).

The banana is a large, herbaceous plant consisting of a branched, underground stem (a *rhizome* in the strict botanical sense, but also often called a *corm*) with roots and vegetative buds, and an erect pseudostem composed of tightly packed leaf bases. The apical meristem, which is located in the centre of the pseudostem at about soil level, gives rise to a succession of leaf primordia. Each primordia grows upwards and differentiates into a leaf base, a petiole and a lamina. The petiole and tightly rolled lamina emerge at the top of the pseudostem in the centre of the cylinder formed by older leaf bases. The leaves emerge at different rates depending on the cultivar and environmental conditions. A banana plant generally has 10-15 functional leaves at flowering and five to ten at harvest, but numbers can be less due to environment and disease constraints. The plant propagates by forming a sucker, which is an outgrowth of a vegetative bud on the rhizome. A young sucker emerging from the ground is called a *peeper*. A number of suckers can arise from each rhizome. Traditional planting material is either the sucker or bit which as a piece of rhizome with a vegetative bud; recently, plants derived from tissue culture have been utilized in commercial operations.

The evolution of edibility in the *Eumusa* series of cultivars most probably began with wild *M. acuminata*, sub-species, which occur naturally in an area stretching from South Asia to Australia, but are concentrated mainly in South East Asia (Simmonds, 1962). Another key event was believed to have been hybridized with *M. balbisiana*, a more drought tolerant species. The genome of wild *M. acuminata* is represented by AAW and wild *M. balbisiana* by BBW. Diploid cultivars derived solely from *M. acuminata* are designated by AA and diploid hybrids with *M. balbisiana* by AB.

Triploidy, another important step forward in the evolution of banana, is thought to have been arisen following the fertilization of viable, diploid egg cells, which can be formed when meiosis breaks down at the second division, with haploid pollen (Simmonds, 1962). Triploid cultivars are found in the AAA and AAB and ABB genomic groups. Only a few, natural cultivars have been recognized as belonging to the AAAA, AAAB, AABB and ABBB genomic sub-groups (Richardson *et al.*, 1965; Shepherd and Ferreira, 1984). These tetraploids, which have robust pseudostems and leaves that tend to droop, are believed to have arisen from the fertilization of triploid egg cells by haploid pollen. Tetraploid hybrids have been artificially bred for disease resistance and are becoming important in some countries.

A number of varieties/cultivars belonging to different genomic groups have been reported from different parts of the world. The same clone can have different names in different locations. This is especially true in Papua New Guinea, a country with over 100 languages, where the names of cultivars can vary between villages. 'Silk', a synonym of West Indian origin, gives its name to the clone that is known in India, as 'Rasthali', in the Philippines as 'Latundan' and in Australia as 'Sugar'. Cultivar names in some countries begin with the local word for banana. For example, names in Malaysia and Indonesia begin with 'Pisang', in Thailand with 'Kluai', in Vietnam with 'Chuoi', in Hawaii with 'Maila' and in Western Samoa with 'Fa' 'i', (For details please refer Jones, 2000).

Total world production of bananas is estimated at over 76 million metric tonnes (FAO, 1993). The (historically) recent rise of the export trade has a major influence over the economics of many nations. Total area under banana cultivation in the world is estimated to be 26.28 lakhs, out of which, India shows 2.82 lakhs hectares. Annual production of banana in India is about 10.4 million tonnes.

Banana cultivation, in different parts of the world has been threatened with extinction and production is severely reduced because of the appearance of several debilitating diseases, which develop to an epidemic level. Diseases caused by viruses contribute to the maximum loss in yield and quality and they are one of the most important limiting factors in the successful cultivation of banana. Descriptions of various viral diseases of banana in different countries have been given by various authors (Simmonds, 1982; Wardlaw 1972; Magnaye and Valmayor, 1995; Singh, 1996; Jones, 2000). The present book deals with all aspects of viral diseases of banana occurring in different parts of the world and upto date information have been included.

1

BANANA BUNCHY TOP

Banana bunchy top is the most important and serious viral disease in most of the tropical countries (Table 1). This disease is limiting or threatening the production in one quarter of the world's banana growing areas. In contrast to most economically important plant viruses, relatively little is known about banana bunchy top virus (BBTV). This was not always the case; in the 1920s, knowledge about BBTV was very much state of the art and the control of the disease in Australia provides one of the most dramatic and successful examples of the plant virus control using phytosanitary methods. Wardlaw (1972), and Dale (1987) and Jones (2000) have given details of the BBTV. Dale (1988, 1990) has suggested the future line of work.

Table 1. Occurrence of banana bunchy top disease

Country	*References*
Pacific	
Australia	Magee 1927
Fiji	Magee, 1927
Guam	Beaver, 1982
Hawaii (USA)	Dietzgen and Thomas, 1991
Kirbati (formally Gilbert Is)	Shanmuganathan, (1980)
Samoa (American)	Magee, 1967, Brooks, 1999
Samoa (Western)	Magee, 1967
Tonga	Magee, 1967
Tavalu (formally Ellice Is)	Campbell, 1926
Wallis Island	Simmonds, 1933

Asia	
China	Diekmann and Putter, 1966
India	Hector, 1925; Padwick, 1940; Nandi, 1941; Magee, 1953
Indonesia	Sulyo *et al.*, 1978
Japan: Bonin Islands	Gadd, 1926
Japan: Okinawa	Kawano and Su, 1993
Malaysia (Sarawak)	Magee, 1953
Pakistan	Khalid *et al.*, 1993; Khalid and Soomro, 1993; Soomro *et al.*, 1992
Philippines	Ocfemia, 1926; Castillo & Martinez, 1961
Sri Lanka	Bryce, 1928
Taiwan	Sun, 1961
Vietnam	Vakili, 1969
Africa	
Burundi	Sebasigari and Stover, 1988
Central African Republic	Diekmann and Putter, 1996
Congo	Wardlaw, 1961
Democratic Republic of Congo (formerly Zaire)	Manser, 1982
Egypt	Magee, 1927; Fahmy, 1924
Gabon	Manser, 1982
Malawi	Kenyon *et al.*, 1997
Rwanda	Sebasigari and Stover, 1988
Somali land	Castellani, 1955
Unconfirmed reports	
Bangladesh	Foure and Manser, 1982
Burma	Buddenhagen, 1968
Hongkong	Buddenhagen, 1968
Kampuchea	Stover, 1972
Laos	Stover, 1972
New Caledonia	Buddenhagen, 1968; Kagi *et al.*, 2001
Sabah (North Borneo)	Magee, 1967
United Arab Republic	Buddenhagen, 1968
Malaya	Magee, 1953
Papua New Guinea	Magee, 1953, Wardlaw, 1961

1.1. GEOGRAPHICAL DISTRIBUTION, INCIDENCE AND ECONOMIC IMPORTANCE

Bunchy top is the most important serious virus disease affecting banana and the only known banana virus not present in the Western Hemisphere. The geographical distribution of BBTV has been presented in Table 1.

In a critical survey of the early history and the spread of the bunchy top disease of bananas, plantains and abaca. Magee (1963) reported how the disease first caused serious losses in an important export industry in the Fiji and references were made about it in the local press in 1889 and in Kew Bulletin in 1890 and 1892. In an official letter from the Governor in 1890 (cited by Magee, 1953) it is stated that Cavendish banana and the local plantains were being attacked and killed off, that the disease spread rapidly to leeward; that by the careless transportation of plants (presumably diseased) it has also spread to windward, that its attacks appeared capricious, in that healthy plants did not necessarily suffer because of their closeness to infected ones; that healthy and affected plants may be in the same mat; that the malady occurred in all kinds of soils and situations and that, no such disease was ever known before. Photographs of the same date and detailed description of the more evident symptoms, leave no doubt as to the identity of the disease.

Fiji

In this group of Islands, where the banana export industry began in 1877 with shipment of 3,100 bunches, production rose to 3,59,000 bunches in the course of ten years, and the peak reached in 1892 with the export of 7,88,000 bunches. During the following three years Bunchy Top resulted in decline, so that in 1895 only 1,14,000 bunches were exported. From 1896 onwards, little improvement was noticed as 4,63,000 bunches were shipped in 1899, but in 1905, production again declined to 1,47,000 bunches. Magee (1953) stated that temporary recovery in 1896, hitherto unexplained may be due to importation and gradual distribution of Gros Michel planting material brought from Jamaica in 1892, this variety being considerably more resistant to the disease than the Cavendish varieties. Magee also recorded that the revival of the banana industry from 1912 to 1916 with export of about 1.5 million bunches per annum, was probably not due to the organized efforts of control or eradication of the diseased material but due to the planting in the new areas. Gros Michel and Veimama (Robusta) did better than Cavendish varieties. The first recorded outbreaks of bunchy top of banana were in Fiji in 1889, though reports and photographs taken around that time leave little doubt that the disease was present there as early as 1879 (Darnell—Smith, 1924; Magee, 1953). A survey by Magee in 1937 (Magee, 1953) revealed that the incidence of BBTV in plantations and gardens was between 5 and 30%. The virus has become entranced in Fiji and still remains a major problem.

Australia

The disease was reported in Australia, in the Tweed River area of New South Wales, in 1913, when a new industry was being established and unfortunately, infected banana suckers were brought in from Fiji (Darnell-Smith, 1919; Darnell-Smith and Tryon, 1923; Magee,1927, 1953). The disease became wide spread with increased area. Magee (1927) recorded that by 1927, in the Tweed and Brunswick districts of New South Wales, profitable banana growing had practically disappeared, the disease being present in all the plantations. The extent of bunchy top disease was 5 to 90% in the two districts. In the Curumbian District in South-East Queensland, only four producing plantations remained in 1925 out of 100 in 1922 (Dale, 1987). In Samoa districts production fell from 37,000 cases in 1958 to none in 1961 due to failure of small growers to co-operate to control bunchy top (Anonymous, 1962). The loss to the individual cultivators was further aggravated by the high prices paid for the land during the period when the industry was prosperous, plantations in bearing having realized £300 per acre and virgin land covered with banana £150 per acre. The usual practice when plantations became unprofitable, was to desert them or to turn in stock to eat down the plants. As the average area of deserted plantations varied from 10-15 acres, several plantations had to be grouped together for profitable dairy enterprises.

By 1927, bunchy top was present along 300 km stretch of the east coast of Australia from Grafton in Central New South Wales to Yandina in southern Queensland. An isolated outbreak at Innisfail, a further 1300 km to the north was detailed in 1924 but subsequently eradicated.

Production reached its peak in New South Wales in 1922, but two years later the industry was in precarious conditions. As a result intensive studies on disease was started and a scheme for control, based on inspection, eradication, re-planting with disease-free material and related ancillary measures, was undertaken jointly by the Department of Agriculture of New South Wales and Queensland and continued for years (Eastwood 1946, 1947). This proved very effective and indeed, may be indicated as one of the outstanding examples of success among phytosanitary procedures. As a result, industry was rehabilitated and the area increased. The bunchy top incidence was greatly reduced to a point where it had practically disappeared (Magee, 1936, 1953; Eastwood, 1947). By the mid-1920's, the disease had spread to Currumbin district in South-East Queensland and South to the Richmond River. An out-break near Innisfail in North Queensland was eradicated (Dale, 1987).

India

In India, unconfirmed reports of the occurrence of bunchy top have been made from Calcutta by Hector (1925), from Cuttack by Padwick (1940) and from Assam by Nandi (1941) but its presence was officially confirmed in 1943 in the Travancore State (Now, Kerala State). The disease has now spread to Andhra Pradesh, Tamil Nadu, Orissa, Maharashtra, Bihar, Karnataka, West Bengal, Assam and Uttar Pradesh (Varasi, 1954; Capoor, 1967; Singh, 1979; Singh, 1988a, 1996). A loss of about Rs. 40,000,000 annually has been reported to be due to this disease, in Kerala State alone (Mehta *et al.*, 1964). Surveys conducted revealed a fast spread of the disease with heavy incidence in Kottayam, Quilon and Trivandrum districts. Currently the disease has spread all over the Kerala State and in the adjoining districts of Kanyakumari of Tamilnadu and Coorg and Mandya in Karnataka. The disease has been reported spreading very fast in parts of Maharashtra (Kamat and Patel, 1951). Mehta *et al.* (1964) reported the disease to be a limiting factor in banana industry. In Karnataka state, the incidence and spread of bunchy top has been observed faster in neglected banana plantations especially when banana is a subsidiary crop in betelvine plantations. In pure, irrigated banana plantations the spread of the disease is negligible (Sharma, 1988). When infected suckers are planted the plants never flower and result in 100% yield loss. When plants of var. Robusta got infected just before flowering, the reduction in finger length, girth and weight were 31.4, 34.6 and 75% less, respectively (Alagiamanavalan *et al.*, 1973). Weight of bunch on infected plants was only 10.5 kg as compared to 31 kg in healthy plants. Time taken for maturity of fingers was 111 days in diseased plants compared to 93 days in healthy plants. In the hills of the lower Palanis in Madurai district in Tamil Nadu where hill banana "Virupakshi" (Syn: Sirumalai, Vella Vazhai), a prized fruit, was grown in more than 10,000 hectares as a rainfed perennial crop, is badly devastated by bunchy top. Kesavamurthy (1980) estimated that more than 60% of this area is completely damaged, about 20% of the area is with about 10-25% incidence and the remaining 20% of the area is with less than 5% incidence. Singh (1989) observed bunchy top symptoms in banana in Nagaland (Unpublished data).

Ceylon

Bryce (1928) considered that bunchy top disease might have reached Ceylon from Fiji. The disease first appeared in 1913 initially in the Colombo district and subsequently it spread to central and Western provinces. It is at present a limiting factor in banana cultivation in certain areas as the varieties in cultivation are susceptible.

Malaya

It is still not certain whether bunchy top disease is present in Malaya. Magee (1953) stated that he has observed a mild chronic form in abaca, of Philippine origin, at Segang, near Kualalumpur.

Indonesia

Banana bunchy top disease was first reported in Bandung, West Java in 1978 (Sulyo *et al.*, 1978). Following this finding, some field observations in the 1980's indicated the occurrence of the disease in Banyawangi, East Java in an abaca Plantation and in other Islands. Now, this disease is known to occur in most banana growing area in West Java especially regencies surrounding Bandung and gradually spread to the Western parts of Central Java.

Philippines

Banana bunchy top disease is widely spread among existing banana cultivars and plantations in the Philippines. This is considered as the most serious disease of the *Musa* family (Roperos and Magnaye, 1991). In 1988, the banana export industry suffered an estimated loss of at least 5 million because of BBTV alone. The disease is present in practically all-commercial plantations in the Philippines. Although its presence has been confirmed not only in Mindanao, but also in other parts of the country such as Palawan, Beguio, Laguna, Bicol, Camiguin Island, and Calapan Mindoro. The extent of infection and production losses has not been definitely quantified.

Taiwan

Banana bunchy top was observed in Central Taiwan as early as the beginning of 19th Century (Sun, 1961), but origin of the out-break is unknown (Hwang, 1991). A cultivar of Taiwan Cavendish Robusta, which was then widely grown in this area, was found more tolerant to this disease. However, in recent years, the variety has also been found to be susceptible. The disease has built up to an epidemic level in Central Taiwan where ratoons are grown. A recent survey revealed that 15-20% of the total banana crop was affected in the region. In the Southern region where annual replanting is a common practice, disease incidence is kept under 1% in commercial banana plantations. Su and Wu (1989) developed a direct ELISA method using monoclonal antibodies to detect BBTV in crude sap from BBTV-infected banana plants. The method is now implemented in tissue culture programme with which 12 million meristem plantlets have been propagated for commercial planting since 1984.

Egypt

According to Fahmy (1924), bunchy top is a serious disease in Dwarf Cavendish banana in Egypt and has been known since 1901. It has limited the extent of banana production.

Other Countries

The disease is not present in Israel. Vakili (1969b) has reported that many varieties of banana are infected on the high plateaus of the Central South Vietnam. This is ascribed to the introduction of abaca, presumably infected, into this region. Johnston (1964) recorded widespread occurrence of the disease in Tonga.

1.2. SYMPTOMATOLOGY

The plants may be infected at any stage of the plant growth. The identification of the initial symptoms, however, is rather more difficult.

External leaf symptoms

The first evidence of bunchy top disease is noticed on the leaves. In a few cases of secondary infection, irregular, dark green, 'Morse Code' streaks about 0.75 mm wide occur along the secondary veins on the underside of the mid rib. At first, one, two or several streaks may be present, usually to be followed by the others in the same region. The abnormal markings may vary from a series of dark green dots to a continuous dark green line, an inch or more in length, with a rigged edge. In some cases, secondary infection symptoms may be perceptible in the newly emerged, or emerging, tightly rolled leaf as pale whitish streaks along with the secondary veins. The leaf may be smaller than usual, slightly chlorotic and with the margin of lamina wavy and slightly rolled upwards (Fig. 1).

Fig. 1

Fig. 2

Fig. 3

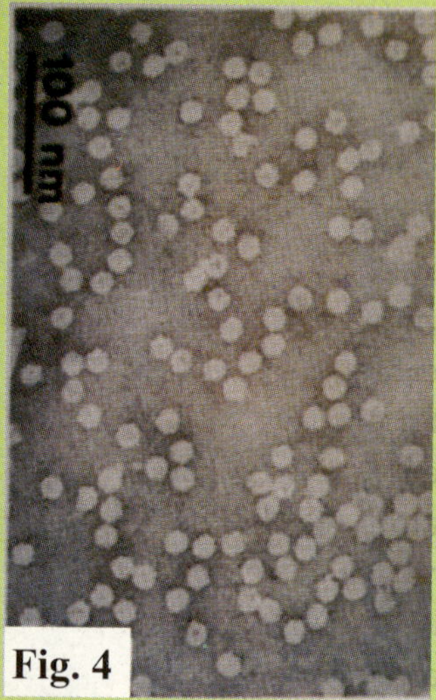

Fig. 4

Fig. 5

The presence of interrupted dark green streaks along the secondary veins of the lamina, or along the mid rib, or petiole, is considered to be the most definite and reliable symptoms of bunchy top.

As the successive leaves of an infected plants emerge, they are seen to be more and more abnormal. Leaves become progressively smaller in size, both in width and length. With limited elongation of petioles, leaves remain abnormally erect. When several abnormal leaves have emerged, an extreme congestion is apparent at the crown, and instead of the normal symmetrical disposition of the leaves, a 'rosette' condition develops (Fig. 2). This characteristic symptoms develops after several weeks of infection. Affected and healthy leaves are different in texture, the petiole, midrib and lamina of the former being harsh, brittle and can be easily snapped when bent or crushed, in marked contrast to the elastic and pliable nature of healthy leaves (Magee, 1927). Bunchy top affected leaves show distinct rigidity and are not readily shaken by winds, they also show an apparent resistance to wilting. However, Magee (1953) has emphasized that symptoms differ with different strains of the virus, varieties and climatic factors, which greatly influence them. At times, symptoms of mixed infection of BBTV, banana streak and infectious chlorosis are seen in the field. Another characteristic symptom is the presence of dark-green streaks on the leaf mid-rib (Fig. 3).

Internal symptoms

A microscopic examination of pseudostem and bulb tissues reveals no pathological changes of diagnostic value. The yellow or brown streaking, *i.e.*, the tannin ducts being characteristic of both healthy and affected plants alike, particularly as they become older. But similar observations of leaf tissue reveal pathological condition. The virus results in a characteristic disorganization of the phloem tissue of the vascular strands. The extent to which vascular strands are altered depends upon the intensity of infection. In severe primary infections, every vascular bundle in the leaf is bound to show some modifications.

Magee (1927, 1939, 1953) has described the following modifications induced by the bunchy top infection in banana :

(*i*) In the phloem, the development of fibrous sheath is suppressed, its place being occupied by cells with many chromatophores.

(*ii*) The ground tissue contiguous with the phloem becomes 'gorged' with chromatophores whereas in normal plants this region is almost devoid of chlorophyll.

(*iii*) The ground parenchyma cells in contrast with the phloem are stimulated to division and form somewhat angular cells by the formation of a thin cellulose partition walls in all planes. Each newly formed cell contains numerous chromatophores and a distinct nucleus.

(*iv*) Similar 'appositional' growth of walls in all planes takes place in the cells of the phloem and gives rise to numerous angular cells along their length. These cells are uninucleate.

Proliferation of the phloem and adjacent tissue has not been observed in the rhizome or roots. Nair *et al.* (1966) investigated the anatomical changes induced in banana var. Nendran as a result of bunchy top virus infection when different levels of CaO and MgO were applied in soil in addition to optimum N, P and K levels. Parenchymatous cells of the sheath undergo diminution in size after the incidence of the disease, such diminution being lowest in the case of diseased plants growing on soils with a CaO/MgO ratio of 3:0. The same is true in xylem vessels and phloem cells of the leaf sheath. The degeneration and dis-organization of vascular bundles and diminution of parenchyma cells of the petiole were minimum where the ratio of CaO/MgO was 3:0. In the case of the root tissues, maximum diameter of parenchyma cells, xylem vessels and phloem cells was maintained by plants growing on soil with CaO/MgO ratio of 3:0 even at an advanced stage of the disease. The endospermal cells and the pericycle of the root also maintained the same trend. The root tissue is comparatively less affected by the secondary infection by bunchy top disease. The diameter of parenchyma and phloem cells of leaf sheath and xylem vessels of diseased plants growing on soil with a CaO/MgO ratio of 3:0 was almost approaching that of healthy plants.

1.3. BIOCHEMICAL CHANGES

Amino acid and amide composition in leaves of bunchy top infected plants has been found slightly different from those of healthy plants (Nair and Wilson, 1970). In healthy leaves, asparagine, aspartic acid, alanine, tyrosine, methionine and isoleucine were detected and in diseased leaves also, except tyrosine, all other amino acids were present. Jose *et al.* (1971) reported higher contents of sugar and phosphorus in fruits of bunchy top infected banana fruits and lower contents of total carbohydrates, nitrogen and calcium in var. Gnalipoovan. Nair and Wilson (1972) recorded higher percentage of carbohydrates, lower percentage of total nitrogen and wider C:N ratio in leaves of bunchy top infected plants than the healthy plants. Zhang *et al.* (1997) found lower giberellic acid and isopentenyla denine contents and higher abscissic acid contents in BBTV infected plants. Mohan and Rao (1986) recorded more of phosphorus content in roots of healthy banana plants than BBTV infected plants.

1.4. ETIOLOGY

Although some studies had suggested an association between the banana aphid and the disease (Darnell-Smith, 1924) it was not until the ground—breaking work of Magee (1927) that the viral nature of the disease was established. In the space of 3 years he established that the causal agent was a virus transmitted by the banana aphid and in the infected material.

The causative agent of BBTV was considered to be possible luteovirus (Matthews, 1982; Thomas and Dietzgen, 1991). This assumption was supported by the work of Dale *et al.* (1986) who isolated BBTV specific double stranded (ds) RNAs which resembled the size of dsRNA isolated from plants infected with barley yellow dwarf and beet western yellows luteoviruses. Wu and Su (1990) later obtained virus-like particles (VLPs) which contained a single-stranded (SS) RNA with a molecular weight of 2.0×10^6 KDa and a single protein with a molecular weight of 21,00 KDa. It was finally demonstrated by Harding *et al.* (1991) that instead of an RNA genome, a small ssDNA is associated with the VLPs and with the banana bunchy top disease. ssRNA were of about 1 Kb and one major protein of M 20100 dsRNA was synthesized from nucleic acid extracts from these particles and cloned. One clone from plants PBT 338, hybridized specifically, (1) with sap extracts from plants infected with banana bunchy top virus but not with sap extracts from healthy plants, (2) with small ssDNA in nucleic acid extracts from infected plants and virus-like particles. These authors have also demonstrated that the occurrence of BBTV associated ssDNA correlates with the transmission of banana bunchy top disease. These results provide strong evidence that these VLPs are the virions of BBTV. This virus contains small ssDNA as its genomic nucleic acid and has a coat protein of M 20100. Thus BBTV is not a member of the luteovirus group as was once suspected (Thomas and Dietzgen, 1991). Further more, they have demonstrated that ssDNA is transmitted with the disease using a probe cloned from the ssDNA. These ssDNA containing VLPs were presumed to the particles of BBTV (Harding *et al.*, 1993).

BBTV is a unique viral species, which contain small ssDNA genome. The virion contains at least three different viral genomic fragments. All these fragments are small ssDNAs with sizes around 1.1 Kb and are probably encapsidated separately into different viral particles as judged by the small size of the virion (19-22 nm). BBTV isolates from Australia, Taiwan, Peoples Republic of China, Tonga, Western Samao and Hawaii are serologically related and have a common etiology (Thomas and Dietzgen, 1991).

1.5. PURIFICATION OF THE VIRUS AND PARTICLE MORPHOLOGY

BBTV has been purified nearly to homogeneity as 18 to 22 nm particles (Fig. 4) (Harding *et al.*, 1991; Thomas and Dietzgen, 1991; Manickam *et al.*, 2001). This virus has also been purified by Wu and Su (1990) and Yeh *et al.* (1994). BBTV has been purified from affected-banana plants independently by two research groups. Iskra *et al.* (1989) detected 28 nm diameter isometric particles in

BBTV-affected plants from Gabon whereas Su & Wu (1988, 1989) isolated 20 to 22 nm diam particles from BBTV—affected plants from Taiwan.

1.6. CLONING, SEQUENCING AND CHARACTERIZATION OF VIRAL GENOME

Harding *et al.* (1993) cloned and sequenced one DNA component of BBTV. This component is present as a circular, ssDNA in the virions and consists of 1111 nucleotides. It contains one large open reading frame (ORF) of 858 nucleotides in the virions sense. This ORF encodes a putative replicase based on the presence of dNTP-binding motif (GGEGKT). Two smaller ORFs (249 and 366 nucleotides) in the complimentary orientation, could not be assigned any obvious function. Neither of these ORFs had significant sequence homology with any known DNA plant virus gene or gene product. Computer analysis of this component predicted a strong stem loop structure in the virion sense putative untranslated; a non-nucleotide sequence in the loop was nearly identical to the non-nucleotide invariant loop sequence of geminivirus and coconut foliar decay virus. BBTV has some characteristics in common with geminivirus but can not be classified as one.

1.7. PROPERTIES OF THE VIRUS

The virus particles are single-stranded DNA with a relative molecular mass (Mr) of 2.0×10^6 and a coat protein sub unit with Mr of 20,000 (Wu and Su, 1990). The virus has an $A_{260/280}$ ratio of 1.46. Component 1 from Hawaii in BBTV has 1110 nucleotides in length and shares 98% nucleotide sequence identity with the BBTV DNA component 1 of the Australian isolate. Components 3 & 4 contain 1057 and 1017 nucleotides in length respectively. The virus possesses a multi component genome, consisting of at least six circular ssDNA components, each c 1000-1100-nucleotides long (Wu, 1997; Wu *et al.*, 1994; Yeh *et al.*, 1999; Burns *et al.*, 1995; Xie and Hu, 1995; Karan *et al.*, 1997; Wanitchakorn *et al.*, 1997).

Yeh *et al.* (1994) characterized the viral genome and identified the viral-associated dsRNA (VADs) component of BBTV. The ssVADs were shown to be the viral genomic DNA and the dsVADs are the likely replication forms. Both dsVADs and ssVADs contain DNA species that cross hybridize to each other. Under the electron microscope, both circular and linear DNAs have been observed from these VAD series. Two over-lapping VAD clones have been isolated and their combined sequences indicate that their corresponding VAD, named as BBTV component 2, is circular with a size of 1095 nt. The largest ORF spans from nucleotides 82 to 390, which is sandwitched by a putative TATA box, 54 nt upstream from the first ATF codon, and a AATAAA motif, 7 nt downstream from the UAA codon. Although the sequences of this VAD is different from the previously reported BBTV component 1 (Harding *et al.*, 1993) and the genome of Coconut foliar decay virus (Rohde *et al.*, 1990), stretches of highly homologous sequences were found between BBTV component 2 and the other two viral genomes by computer analysis. The sequence of BBTV component 1 is highly homologous to one of the clones, PBTs 26, in the VAD library. There are still many clones, which hybridize with SS viral genomic DNAs but not with PBTs 26 and BBTV component 2, indicating that the viral genome consists of more than two components. BBTV probably belongs to a new viral group of plant viruses.

Xie *et al.* (1994, 1995) and Wanitchakorn *et al.* (1997) reported that BBTV has a genome consisting of at least six single-stranded, circular DNA components about 1 kd each. Two cDNA clones representing the full length of two of BBTV genome component from Hawaii were sequenced.-One of the two components encodes a putative replicase gene, which shares 98% sequence identity with the Australian BBTV isolate. Wu *et al.* (1994) cloned and sequenced two circular ssDNA 2 designated DNA I and II of the BBTV. The genome of BBTV consists of 1,106 nucleotides and contains four open reading frames (ORFs) and potentially encoding proteins with molecular masses ranging from 5.4 to 33.18 KDa and one putative stem loop structure of 29 nucleotides. The sequence of BBTV DNAII has 1096 nucleotides and contains four ORFs encoding putative proteins with sizes ranging from 5.6 to 32.77 KDa and one putative stem-loop structure of 32 nucleotides. The largest ORF (Ve) in DNA II encoded a putative replicase with the nucleotide triphosphate binding monif (GGEGKS). DNAs I and II have a 54.3% identity in total sequence and two conserved regions with >80% sequence identity. The two BBTV sequences showed no close relationship with geminivirus DNA sequences but some similarity with the coconut foliar decay virus (CFDV) DNA sequence (about 39% identity DNAs Il & 1 showed considerable with 50.08% and 52.9% identity, respectively). The loop sequences of putative stem-loop structures in both BBTV DNAs were highly homologous and also had considerable similarity with those of geminivirus, CFPV, DNAs and BBTV component 1.

Wanitchakorn *et al.* (1997) reported that a protein of approximately 20 kDa was associated with purified virions and is assumed to be the viral coat protein. The M terminus of this, when sequenced and compared with the predicted amino acid sequence of the large ORF of BBTV DNA-1 to 6, suggested the ORF of BBTV DNA-3, potentially encodes a protein of 19.3 kDa and is the coat protein gene of BBTV.

Karan *et al.* (1994) cloned BBTV DNA component -1 from isolates from different countries and sequenced. This analysis indicated two groups; the South Pacific Group (isolates from Australia, Burundi, Egypt, Fiji, India, Tonga and Western Samoa) and the Asian group (isolates from the philippines, Taiwan and Vietnam). The mean sequence difference within each group was 1.9 to 3.0% and between isolates from the two groups was 10%.

He Zifu *et al.* (2000) cloned and sequenced the DNA component 6 of a Chinese isolate of BBTV. It is 1081 nucleotide in length. Between the component sequenced of Chinese and Australian BBTV isolates there is 85.5% nucleotide sequence identifies in full length, 92.5% nucleotide identity in predicted open reading frame (ORF) and 94.8% amino acid sequence identity of potentially encoded proteins.

La Ping *et al.* (2000) cloned and sequenced BBTV component I (about 500 bp) and found that this fragment is full length component 1 of BBTV. Sequence comparison revealed that the component 1 of Zhangzhou isolate shared homologies of 89.6% and 96% those of the Australian and Chinese Taiwan isolates respectively.

1.8. VIRUS STRAINS

Strains of BBTV are not well defined on the basis of similarity of biological properties, abaca bunchy top virus (ABTV) is very probably a strain of BBTV. ABTV causes a yellow-type disease in abaca (*Musa textilis*) and is transmitted in a persistent manner by *Pentalonia nigronervosa* but is not transmissible by sap (Ocfemia, 1926, 1930). BBTV has been shown to infect abaca (Magee, 1927) causing similar symptoms of ABTV, but, ABTV does not infect banana (Ocfemia and Buhay, 1934). However, the exact relationship between these two viruses is not known. Magee (1948) described a number of Veimama plants in Australia, which showed very mild symptoms of BBTV infection. Suckers derived from these plants could be super infected with an Australian isolate of BBTV but only when inoculated with three times as many infective aphids as usually required, and then only 40% of plants developed normal BBTV symptoms. Magee was unable to transmit BBTV from the mildly infected Veimama plants. The phenomenon of recovery of the Veimama cultivar has also been observed in Fiji (Magee, 1948). The most probable explanation of this phenomenon is the presence of a mild or attenuated strain of BBTV (Magee, 1967; Shimmonds, 1966). Two broad groups of isolates have been identified, based on nucleotide sequence difference between some, possibly all, of the six recognized genome components (Karan *et al.*, 1994). The 'South Pacific' group comprises isolates from Australia, Fiji, Western Samoa, Tonga, Burundi, Egypt and India, while 'Asian group' comprises isolates from the Philippines, Taiwan and Vietnam. These differences are present throughout the genomes of components 1 and 6, but are most striking in the untranslated major common region. No biological differences have been associated with these sequences differences.

1.9. TRANSMISSION

Banana bunchy top virus is not transmitted by mechanical means. Magee (1927) showed that the banana aphid, *Pentalonia nigronervosa* was the major vector (Fig. 5). The virus may be transmitted by infective aphids in 1.5-2 hours or more of feeding on healthy plants along with retarded growth rate leads to a fall in the availability of the virus (Stover, 1972). By daily transfer to fresh plants the virus may be retained in the aphids upto 13 days after the removal from infected plants. Infective power in the aphid after feeding on disease plants develops in from a few hours to about 2 days (Magee, 1940). Studies in Taiwan (Sun, 1961) showed that aphids feeding for a period of 2-96 hours on diseased plants increased their transmitting ability from 20 - 100%. Ocfemia and Buhay (1934) reported that aphids required between 24 and 48 hours after feeding on infected abaca for the effective transmission of the virus. Magee (1940) observed that bunchy top was transmitted both by winged and non-winged adults. Nymphs were more efficient than adults. He worked out the minimum feeding period for acquiring the virus by the nymphs to be 17 hours but better results were obtained with 24 hours of feeding. To transmit the virus, 1.5 to 2 hours feeding was needed on susceptible test plants. The virus was not transmitted transovarially. Temperature of 10°C and 15°C lessen the utility of infective aphids and their inclination to feed on healthy plants. Khurana (1971) failed to transmit the virus when 15-25 aphids/plants were used for transmission studies. Bunchy top virus has also been transmitted experimentally by *Myzus persicae*, *Aphis gossypii* and *A. spiraecola* from banana to few herbacious host (Rao, 1980 a, b). However, this needs confirmation. The virus is mainly transmitted by infected suckers. Rao (1980 a, b) has also transmitted the virus by leaf patch grafting from banana to *Catharanthus roseus*. Banana bunchy top was successfully transmitted to the healthy banana plants by core and patch grafting (Sastry *et al.*, 1980). Inoculations to healthy suckers were made by removing a core of tissue (1.5 to 3 cm diam and 2-3 cm in length) with cork borer and replacing with similar sized core of bunchy top infected rhizome and tied them with polythene tape. In the patch grafting method about 5 cm rhizome portion was removed from the periphery (top to down) with a sharp knife and same size of infected rhizome was kept together and tied with polythene tape and potted. The grafted plants show bunchy top symptoms 30—40 days after graft inoculation.

Thomas *et al.* (1995) reported that BBTV was also transmitted through tissue culture.

1.10. SPREAD OF BUNCHY TOP

Allen (1978a, b) studied the spread of bunchy top in time and space and also worked the efficacy of roguing in eliminating the disease. He worked the relative infection rates and eradication efficiency in different situations. The values determined for relative infection rate and eradication efficiency in one disease outbreak were used to predict the course of disease control in a separate outbreak (co-efficient of determination, 0.7). They have also been used to simulate various disease control strategies in which the ratio of N_2/N_1, was used a measure of control efficiency (Allen, 1983). A ratio of > 1 indicates that control is failing while a ratio of < 1 indicates progress towards eradication. When ratio equals 1, other parameters are to be looked into. By using the equations given (Allen 1978 b), the plants that are likely to be infected can then be destroyed before they become new sources of bunchy top. The probability of an aphid flying more than 100 m is < 0.001 and such an isolation distance has proved effective in practice (Allen and Barnier, 1977). Incubation period, relative infection rate, detection efficiency and eradication efficiency have been dealt in detail.

1.11. HOST RANGE AND VARIETAL SUSCEPTIBILITY

The host range of BBTV seems to be confined to six species of the genus *Musa*. These include *Musa ensete, M. sapientum, M. sinensis, M. paradisiaca, M. textilis and M. acuminata* ssp. *banksii* (Manser, 1982). There is considerable confusion surrounding the taxonomy of the *Musa* genus because of the high incidence of polyploidy and therefore speciation is different, if not impossible to define. Species in several closely related plant families, including the Araceae (*Alocasia* spp), *Calladium* spp, *Dieffenbachia* spp., *Xanthosoma* spp.; Cannaceae (*Canna* spp); Heliconiaceae (*Heliconia* spp); Strelitzeaceae (*Strelitzia* spp.) and Zingiberaceae (*Alpinia* spp.), *Costus* spp, *Hedychium* spp. are also colonized (Wardlaw, 1961). Ram and

Summanwar (1984) reported that *Colocasia esculenta* carry BBTV symptomlessly. *Heliconia* sp. has been implied as BBTV reservoir based on bunchy top-like symptoms in the field, but this has not been confirmed by transmission experiments. Rao (1980a) reported that BBTV can infect following plants when inoculated artificially through aphid vector. The symptoms produced on these hosts are given below, however, this is not confirmed.

1. *Cucumis sativus* L. — Stunting
2. *C. melo* — Stunting
3. *Gynura aurantiaca* — Vein clearing & yellowing
4. *Catharanthus roseus* — Yellowing

No species or variety of the genus *Musa* is known to be immune or resistant to bunchy top. However, from Vietnam, Vakili (1969 a) has reported *M. balbisiana* and some wild *M. acuminata* as resistant and ABB *Musa* clones to have a high degree of field resistance. Gross Michel is more resistant to infection than the Cavendish group of varieties (Magee, 1948). Magee (1948) considered that the introduction of "Gros Michel" to Fiji in the early 1900s contributed to partial rehabilitation of the bunchy-top devastated industry. Cultivars within the Cavendish sub-group form the basis of the international banana export trade and are generally susceptible to BBTV. However, it appears that not all cultivars with an AAA genome are similarly susceptible. Vakili (1969 a) reported *Canna* with possible bunchy top virus symptoms but Magee (1927) was unable to transmit banana bunchy top to *Canna* or any other genus except *Musa*. Yang (1970) reported that out of 35 varieties tested, only Lun-Ya-Chiao was found resistant. Chinese dwarf and Hsien-jenchiao are highly susceptible. All the 45 cultivars grown in India have been reported susceptible (Capoor, 1967), but Jose (1981) has observed cultivars Kanchikela and Vennattukunnan as resistant and Karpooravally, Karimonthan, Ambalakadali, Sennachenkadali, and Vadakkankadali as tolerant under artificial inoculations. High degree of resistance was observed in Gros Michel and Veimama bananas grown in Fiji. Resistance in Gros Michel was either physiological or structural, but once it contracts the disease, its reaction was as severe as that of Cavendish group. The resistance in Veimama banana is due to acquired immunity.

1.12. ECOLOGICAL RELATIONSHIP

Bunchy top disease is known to occur under tropical, sub-tropical and temperate conditions, indicating thereby that climatic factors exercise little influence either on the incidence or development. On the other hand, seasonal variations show marked effect. In Australia, disease incidence is low during winter months, whereas in warm months of January to March, when aphids are most plentiful, the disease reaches its peak. A particular sequence of weather conditions such as period of rain following a dry spell during summer, also results in rapid intensification of existing infections. Topography plays an important role. Bunchy top is usually more on the lower, more sheltered moist and warmer areas of hillside and less near the exposed brow of the hill in Australia. In the former cases, plant growth is more vigorous and aphids are more abundant. In more fertile soils, banana showing more vigorous growth, are generally more susceptible. Kolkaila and Soliman (1954) observed that in Egypt, aphid activity was greatest during winter months (December-January) with a temperature of 20-22°C and a relative humidity of 60-70%. The alatae forms appeared during the rainy period. As the temperature rose and the humidity fell during March-April, there was a reduction in the alatae population. Some alatae forms appeared during summer also and they were responsible for the incidence of the disease in September-October. Varma and Capoor (1958) observed that the aphids bred throughout the year in banana but were more common during July. Menon and Christudas (1967) stated that climatic conditions existing during the summer months and rainy weather are unfavourable for the banana aphid in Kerala state. Aphid infestation is high during August to February when the temperature and rainfall are moderate. They also observed that aphid population was higher on bunchy top affected plants than on healthy plants. Six varieties *viz.*, Conoor, Nendran,

Ennabaniyan, Boothibali, Peyan, Koath Bontha and Kali were more susceptible to aphid infestation than the other 24 varieties tested. For the downward movement of the virus after inoculation with the aphid, a minimum time of 5 days and maximum of 10-15 days has been reported depending upon the vigour of the plant (Sam Raj *et al.*, 1970). When movement starts, it takes less than a day to traverse pseudostem.

While investigating the effect of host nutrition on bunchy top disease, Nambiar and Nair (1965) reported that soil samples from the base of diseased plants were more acidic and had a higher content of organic matter, N and available P but a lower content of total K and Ca, with very little Mg. In infected leaves, there were relatively large amounts of N and P and very high content of K. Application of time with NPK had little effect on the incidence of the disease while Mg alone or in combination with Ca exerts an appreciable influence on the appearance of symptom. Nair and George (1966) reported that the CaO/MgO ratio of 3.71 in the leaf showed no significant correlation with disease incidence or resistance. However, Nair and Pillai (1966) reported higher concentrations of N, P and K in infected leaves than in healthy ones. Plants, which delayed the symptom appearance, had higher contents of Ca and Mg as compared to those plants, which exhibited earlier infection. They also concluded that Ca or Mg alone did not have any effect on delaying the disease appearance. Higher CaO/MgO ratio seemed to have a role in imparting resistance to the plant. They further pointed out that CaO/MgO ratio of 3.5 to 4 or a CaO + MgO/K_2O ratio of one or near one in the plant tissue can successfully delay the appearance of symptoms until the emergence of bunch. Pillai and Nair (1974) while continuing their investigations on the effect of nutrition on bunchy top disease, concluded that CaO + MgO/K_2O in the plant tissue exerts some influence on the resistance of the tissue to infection by the virus. Nair *et al.* (1966) have also reported reduced anatomical changes in tissues of diseased plants grown in soil having CaO/MgO ratio of 3.0.

Kesavamurthy (1980) observed that in the hills of lower palanis in Madurai district of Tamil Nadu, tremendous changes in the ecosystem have taken place in the seventies and the trend indicate a continuance of these changes in the future too. Hill banana (var. Virupakshi) grown in this region in more than 10,000 hectares as a rainfed crop at the altitude of 750-1800 metres, is badly devastated by bunchy top. In this region, in fiftees, banana and coffee existed side by side with cardamom also in this range on a significant level. More than 2000 hectares of cardamom then is now reduced to less than one fifth of that area. About 90% of this one fifth of the area converted were planted with banana. Most of this happened in the sixties and early seventies. Monocropped banana and banana intercropped with coffee was edged out by diseases and pests, most important of which is the bunchy top. Perennial nature of hill banana cultivation, contiguous fields over hills and valleys of this vast banana tract and the humid climatic conditions favoured the multiplication of the vector aphids and the windy months helped the aphids to spread fast.

Diseased suckers play a key role in the epidemiology of BBTV. The secondary spread occurs through *P. nigronervosa*. Dispersal nationally and over long distances internationally is an infected material and the virus has probably been introduced to most countries this way. Dispersal by aphids is essentially confined to short distance spread.

Allen (1978b) showed that in two case studies, the mean distance of new infections from their source of inoculum in an established plantation was 17.2 m, about two thirds of new infections were within 20 m, and all but about 1% were within 8.6 m. Allen and Barnier (1977) found that a new plantation (planted with healthy planting material) had a greater than 88% chance of becoming infected within 12 months if it was adjacent to an infected plantation, that this dropped to a 27% chance if the new Plantation was 50-1000 m distant and none of 19 new plantations that were more than 1000 m from an infected plantation became infected within 12 months.

Natural infection of BBTV in plant species outside *Musa*ceae has not been conclusively identified although Ram and Summanwar (1984) have reported that *Colocasia esculenta* as a symptomless carrier of the BBTV. This may have important epidemiological significance as this species is wide spread in many areas where the virus occurs and has been reported as a natural host of *P. nigronervosa* (Mehta *et al.*, 1964), the aphid vector of BBTV. Therefore, it is essential to confirm that *Colocasia esculenta* is a host and whether it is a natural reservoir of the virus. Other non-*Musa* species such as *Canna* sp. (Vakili, 1969a) and *Heliconia* sp. (Magee, 1967) have been implied as virus reservoir based on bunchy top-like symptoms in the field, but these have not been confirmed by transmission experiments. The three important regions on the plant where the colonies concentrate are : (*a*) The base of the pseudostem, (*b*) The top region of the pseudostem and (*c*) The leaf axils. Most of the aphids are confined around the base of the pseudostem near the soil level and at times several inches below the ground level. Here the colonies are sheltered by the old dried up leaf sheaths or soil. On the top of pseudostem, aphid colonies remain sheltered by the dried up outer sheaths. Colonies are found in the leaf axils also. But during the monsoon the leaf axil gets filled up with rain water and aphids seldom thrive in the leaf axil. During seasons when aphids are abundant, they are seen on parts of plants other than those mentioned above. In some cases aphid colonies are found on exposed surface of tender leaves, mostly on the underside, on the sides of mid rib, or at the tip and within un-opened leaves. On a rough estimate it has been seen that 45% of the total number of aphids present on a plant are confined to the base of the pseudostem, 35% to the top of pseudostem and the remaining 20% to the leaf axil and other regions (Johnson, 1963). Ocfemia *et al.* (1947) found the aphids colonizing on the roots and corms of Manila hemp in Philippines during the dry season. Ogilvic (1927) located aphid colonies on the flowers and tips of fruits, of banana in Egypt. Magee (1927) however, could not find the vector on the roots of banana.

Presence of aphid colonies may not be uniform on all the plant parts in a particular locality. Often one could observe comparatively aphid-free plants in close proximity to plant harbouring large colonies of it.

1.13. DISSEMINATION OF THE APHID

Magee (1927) and Goddard (1929) based on careful observations made in the field, concluded that the dispersion of *P. nigronervosa* is accomplished by flight. But contrary to this observation, Ocfemia (1930) found that aphid is not capable of prolonged flight and that it is spread from plant to plant chiefly by ants. Johnson (1963) made some observations in Kerala state and concluded that the dispersion of aphids takes place more by contact transfer from plant to plant than through any other external agency. Aphids were never seen flying in the field; winged aphids were more sluggish in their movements as their wingless counterparts. Magee (1927) summarized the possible ways in which aphid dissemination would take place.

(1) By contact transfer from plant to plant in the some or adjacent rows.
(2) By the movement of the aphids over the soil.
(3) By the flight of the winged individuals.
(4) By medium to high wind blows.
(5) By the movement of aphid infected suckers; and or
(6) By the translocation of aphid infested soil on the tools and boots of the workers and also on clothes of workers during cultural and harvesting operations.

In the BBTV infected banana-growing areas of Australia, wild or uncultivated musaceous hosts of the virus are very few in number and are probably of little consequences in the epidemiology of the disease. However, in the more tropical growing areas of South-East Asia and the Pacific, such musaceous alternate hosts are much more prevalent and may be a significant reservoir of the BBTV.

1.14. ORIGIN AND BIOLOGY OF THE APHID VECTOR *P. NIGRONERVOSA*

The aphid, *P. nigronervosa* is supposed to be native of South Asia (Stary, 1966) or more precisely to South East (Voelkl *et al.*, 1990) but is now found in the tropics and sub-tropics all over the world.

Works of Froggatt (1923), Magee (1927), Zeck and Eastwood (1929), Smith (1937) and Kolkaila and Soliman (1954) provide some information on the biology of the insect in other countries. In India, Johnson (1963) and Menon and Christudas (1967) have worked on the life history and population dynamics of *P. nigronervosa* in Kerala State (India).

1.15. LOCATION OF THE APHID ON THE HOST PLANT

P. nigronervosa occurs on the banana plants usually in small to very large colonies (Fig. 5). Both winged and wingless viviparous females and nymphs are present in these colonies, the wingless forms being more abundant than the winged forms. On the plants, the aphid occurs mainly in sheltered situations; only very rarely the pests are found exposed.

1.16. LIFE HISTORY

Males of *P. nigronervosa* have not been observed so far. The female breeds parthenogenetically. Egg stage also is not known. The female give birth to young ones. These viviparous females may be winged or wingless (Johnson, 1963). Description of these two forms is described earlier (Zeck and Eastwood, 1929). These two forms have been known to remain in air for over four hours. Generally one aphid produces 32 to 50 young ones during its lifetime, the average being 40.5. One to three nymphs are brought forth per day by a single aphid. From time of its birth as a nymph to the final nymphal moult the insect takes a period of 8-9 days. The longevity of the adult aphid varies from 23 to 37 days.

1.17. HOST RANGE OF THE VECTOR

In addition of banana, the aphid also breeds and infests in nature on several other monocots including *Ravenalia* sp., *Strelitiza* sp. (Magee, 1927), wild banana, *Heliconia behai*, *Heliconia* sp. (Wardlaw, 1972), *H. latispatha*, *Xanthosoma roseum* (Evers, 1968); Tannia (Wolcott, 1933), Calla flowers, *Hedychium coronarium* (Ocfemia and Buhay, 1934), *Languas speciosa* (Sun, 1961), small cardamom (Uppal *et al.*, 1945; Varma and Capoor, 1958); large cardamom (Vasudeva, 1956),

Colocasia sp. and *Alocasia* sp. (Marathe *et al.*, 1982). *P. nigronervosa* var. *caladii* has also been reported on *Colocasia* sp. (Siddappaji and Reddy, 1972), but none of these hosts is infected by the BBTV except *Colocasia* sp. which has been reported as symptomless carrier of BBTV (Ram and Summanwar, 1984). *P. nigronervosa* is also found on *Zingiber officinale, Curcuma clomestica* (*C. longa*) and *Colocasia esculenta* (Zhou Zhong Ju *et al.*, 1995).

1.18. PHENOMENON OF LATENCY, PARTIAL RECOVERY AND MILD CHRONIC INFECTION

In eradication campaigns, one of the most disconcerning aspects is that the disease may remain 'dormant' *i.e.*, latent, in the plant for a considerable time. This phenomenon was first noticed in 1941 and has since been confirmed (Eastwood, 1947). It had, at first, been thought that latency would persist only for two growing seasons (about eighteen months), but experience has shown that it may continue for an indefinite period. Accordingly, the total elimination of the disease is difficult to achieve, unless infected plantations are completely eradicated.

Several reports from Fiji and Australia emphasize the sporadic distribution of the diseased stools in plantations and gardens and the persistence of healthy or symptomless plants in close proximity to severely affected ones. So far, absolute recovery, from a virus disease in plants is unknown. But there may be recovery to seemingly normal growth and absence of symptoms. Such partially recovered plants acquire resistance to the virus and healthy plants in their proximity do not contract the disease. So the virus may exists not only in severe form but also in mild chronic form of the disease may supervene on an initial attack, may develop the green-streak symptoms much reduced in intensity in some leaves only, but not the dwarfing, marginal chlorosis, curling of leaves, rosetting, chocking and reduced size of bunches as in the virulent form. In experiments using a standard transmission/inoculation procedure, Magee (1948) investigated the infection relationship in 'partially recovered' plants of the Veimama variety. Bhakthavatsalu *et al.* (1980) observed that perennial banana cultivation in Thanjavur area in Tamil Nadu passed through the phase of initial acute infection of BBTV and now appear to have acquired wild strains protection. Groves with mild strain is perhaps widespread in this belt and may tend to replace the severe strain altogether. Probably the presence of mild strain alone has permitted Thanjavur's survival as a banana producing area for it is impossible otherwise to explain the fact that the severe type of bunchy top is not nearly so common in this area as would be expected. Mild strain of the virus affecting cv. Nendran (French plantain-AAB) which though symptomless externally, reduced fruit size and crop productivity. However, different isolates of bunchy top virus collected from different places and varieties, when cross inoculated to different varieties did not exhibit the mild or severe differentiation (Anonymous, 1984). Simmonds (1959) pointed out that on the basis of mild and severe infection relationship, there is possibility of controlling the disease. However, the possibility of these mild strains in some cultivars, to be very severe cannot be ruled out (Simmonds, 1966). Magee considers that the entry of bunchy top virus into North Borneo, where it occurs now in banana and abaca can be ascribed to the introduction of seemingly healthy planting materials which, however, carried the mild form of the virus. Moreover, as pointed out by Simmonds (1962), the mild form of bunchy top is not transmissible, the banana suckers it would presumably have to be bulked up from suckers selected in diseased fields.

1.19. DETECTION OF BBTV

Bunchy top of banana can be detected by following methods :

Symptomatology.

Transmission by aphid.

Serology.

Nucleic acid-based assay.

Polyclonal and monoclonal antibodies are now routinely used in ELISA to detect BBTV in field, tissue culture plants and can detect the virus in single viruliferous aphid (Wu and Su, 1990b; Dietzgen and Thomas, 1991; Thomas *et al.*, 1991; Thomas and Dietzgen, 1991;

Thomas *et al.*, 1995; Othman *et al.*, 1996). Triple antibody sandwich—ELISA is optimum assay format for routine virus indexing (Geering and Thomas, 1996).

Various nucleic acid-based assays have been applied to the detection of BBTV in plant tissue and viruliferous aphids, including DNA and RNA probes, labeled either nonradioactively or with 32P (Hafner *et al.*, 1995; Xie and Hu, 1995; Sadik and El-Karim, 1997). PCR has proved to be about 1000 times more sensitive than ELISA or dot blots with DNA probes (Xie and Hu, 1995). Substances in banana sap inhibitory to PCR can be circumvented by sample extraction procedures (Thompson and Dietzgen, 1995) or by immunocapture PCR (IC—PCR).

1.20. DETECTION OF LATENT INFECTION IN PLANTING MATERIAL

Successful control of banana bunchy top in Australia, has been achieved through inspection, eradication and planting with healthy suckers. But the latent infection in banana suckers results in the introduction of this serious disease in new plantations. Summanwar and Marathe (1982) recommended the use of 1% solution of 2,3, 5-triphenyl tetrazolium chloride (TTC) for detecting the latent infection of bunchy top and infectious chlorosis (mosaic) in banana suckers. When sections from parenchymatous tissues from suckers from roots, daughter suckers, swollen buds, mid-ribs and leaves are immersed in TTC solution at 32-35°C, the sections from bunchy top infected plants turn red, from infectious chlorosis, black and colourless from healthy plants within $1\text{-}1^1/_2$ hours. They also stated that any part of the plant can be taken for diagnostic test but they preferred the sections from parenchymatous tissues of suckers, though sections from this tissues takes longer time than from other plant parts for the colour to develop. However, the test did not give consistent results and is erratic and cannot be truly adopted for the disease diagnosis (Sharma, 1988; Singh 1988a). Sharma (1985) used the section from different parts of the suckers, roots, secondary suckers, pith of pseudostem, pseudostem and mid ribs from bunchy top infected infectious chlorosis infected and healthy plants in 1, 2 and 3% solutions of 2, 3, 5-triphenyl tetrazolium chloride at room temperature (24 + 2°C) and also at 34°C in an incubator. Sections from all the categories of plants turned brick red. Colour development was quickest and brightest in sections taken from pith of pseudostem and growing points of secondary suckers. In a mature sucker, colour development was in discending order from the top to bottom of suckers when they were split open longitudinally and sections were taken from different point starting from growth point. Such variations in colour developments were not recorded when sections were taken in horizontal planes. Sections from pith of pseudostem became brick red within 3 minutes at 34°C and within 5 minutes at room temperature irrespective of diseased or healthy plants. Sections from top, middle and base of suckers of pith of healthy and diseased plants developed brick red colour at room temperature within 15, 50 and 90 minutes, respectively. Tender root tips of both healthy and diseased plants gave brick red colour within 10 minutes at room temperature whereas all the sections from old roots did not turn red even upto 90 minutes. Intensity of the colour was maximum and time taken minimum when sections were taken from the pith of pseudostem towards the base and colour intensity decreased and time taken increased when sections were taken towards the top of the pseudostem. It is evident from the above results that staining technique with 2, 3, 5-triphenyl tetrazolium chloride for the detection of BBTV (Summanwar and Marathe, 1982) is unsuitable and unreliable. Unsuitability of this test has also been reported from Kerala for the detection of BBTV (Anonymous, 1985) and from Maharashtra for infectious chlorosis detection (Anonymous, 1984, 1985).

1.21. MANAGEMENT OF BBTV

In severe outbreaks of bunchy top disease, there are two major problems: (1) To secure commercial control in the affected plantation areas and (2) To avoid outbreaks in plantations not yet affected. Experience has shown that, of the various control measures that might be suggested, only exclusion and eradication have proved effective. Remedial measures based on improved cultivation, application of manures, fertilizers, etc., may modify the manifestations of the symptoms but they do not control or eliminate the disease. Since, all cultivated and even wild species of *Musa* are apparently susceptible, breeding for disease resistance has not so far offered very much hope, though differences in susceptibility have been observed. The part played by banana aphid is of paramount importance; if the aphid population should be eliminated by the application of sprays, dusts, etc., and of all diseased plants are destroyed as soon as symptom appear, so that they are not used as planting material, the disease could quickly be reduced to negligible proportions. The various approaches to control this serious malady are discussed below.

1.21.1. Exclusion, eradication and quarantine

The disease is spread in planting material and from plant to plant by banana aphid. Legislation should be introduced to prevent planting material being taken from plantations affected by bunchy top and to require banana growers to destroy any diseased plants. Growers should co-operate to organize an efficient system for detecting and destroying diseased plants. The following methods should be practiced :

(*a*) Prohibition of transportation of any vegetative parts of banana plant or any other member of the genus *Musa* from diseased to healthy plantations.

(*b*) Using only certified suckers for new plantations.

(*c*) No person is allowed to trade in suckers unless special official permit is granted.

(*d*) The immediate destruction and further prohibition of all banana plants and *Musa* sp. in backyard or similar gardens.

(*e*) No transportation of banana fruits from affected areas.

(*f*) Regular surveys by the most competent inspectors in new and old plantations.

(*g*) Education of growers so that they may identify and deal with the disease in its earliest stages. Exclusion of the disease by the measures listed above and complete destruction of diseased stools are considered to offer greater possibilities of success than schemes based on the control of insect vectors. Several workers from different parts of the world have recommended the legislative measures of sanitation for controlling bunchy top virus (Colborne, 1953; Nair and Menon, 1965; Magee, 1967; Buddenhagen, 1968; Morgan, 1969; Graham and Navarathnam,1970; Fireman, 1970; Wardlaw, 1972; Sastry *et al.*, 1980).

1.21.2. Control of aphid vectors

Several workers have tried various chemicals in different formulations to control the bunchy top vectors, *Pentalonia nigronervosa.* Eastwood (1933) recommended spraying of healthy plants with kerosene emulsion to destroy the aphids and treatment of the diseased stools with undiluted kerosene for destroying both the plants as well as the aphids. Hassan (1958) observed that sprays of parathion, malathion, gusathion, diazinon, chlorothion and ekatin gave effective control of the insect. Johnson (1965) have evaluated the relative efficacy of some systemic insecticides and found parathion superior to others. Fortnightly sprays of endrin (0.02%) or diazinon (0.03%) have also been found to reduce the aphids significantly (Menon and Christudas, 1966). Nair *et al.* (1973) evaluated Thiodemeton (solvirex, disyston), Lindane, Dimethoate and Phorate granules @ 50 g and 25 g as plant base and leaf axil application, respectively and observed all insecticides highly effective in controlling the vector except basal application of lindane. They further evaluated these insecticides @ 25 g and 12.5 g/plant of the base and leaf axil application. Disyston as leaf axil application was phytotoxic. All other treatments were highly effective. Regupathy and Kulasekaran (1980) have observed that injecting systemic chemicals like monocrotophos or methyl demeton at 0.05 to 0.2 ml ai plant 3-4 week intervals reduced the vector population considerably. Regupathy *et al.* (1983) also reported methyl demeton as the best treatment among the five insecticides tested. Residual toxicity among the five insecticides has been reported from 14-35 days when aphids were confined to the treated banana plants (Suresh and Regupathy, 1985). The toxicants having residual toxicity from highest to lowest were carbofuran, carbosulfan, monocrotophos, methyl demeton, phosphomidon and dimethoate.

1.21.3. Thermo and Chemotherapy

Attempts made to control bunchy top virus by applying aureomycin have been unsuccessful (Wardlaw, 1972). Application of giberellic acid to infected plants also resulted in about 80% elongation of stem but no differences were recorded in leaf symptoms (Ramaswamy *et al.*, 1966; Govindaswamy *et al.*, 1977). Spraying, wick feeding or dipping of suckers in 500-1000 ppm solutions of tetracycline hydrochloride did not result any recovery of diseased plants (Sastry *et al.*, 1980; Nayar, 1984). Dipping bunchy top infected suckers in 2% benomyl also gave negative results (Sastry *et al.*, 1980). Heat therapy of infected suckers at 40—55°C for different periods did not eliminate the virus (Ramaswamy, 1967; Govindaswamy *et al.*, 1977; Sastry *et al.*, 1980). Nayar (1984) reported 10% success in eliminating the disease by dipping the suckers in hot water at 70°C for 120 minutes.

1.21.4. Eradication of stools

Mechanical eradication of stools have proved quite satisfactory in Republic of Honduras and Central America. Plants and suckers which have previously been effectively sprayed with aphidicides when cut down to soil level and sprayed with heavy gal-oil @ 2 pints/mat also results in complete destruction of stools (Wardlaw, 1972). Use of kerosine and 2, 4-D has also been in practice for eradication of the stools. Beaver (1982) has recommended the use of herbicide, picloram (Keipin formulation) by inserting the toothpick like wooden pins impregnated each with 6 mg of picloram. One to six pins are inserted into each pseudostem and mats are killed within a 21-day period. A second cycle of chemical is necessary if re-growth of suckers arise from the infected mat. Fernoxone has been found very effective for killing the banana plants. Combination of Fernoxone and Methyl demeton has proved very effective for control of aphids as well as killing of plants (Regupathy and Subramanian,1980).

1.21.5. Cross-protection

Cross-protection technique may be one possibility for the control of BBTV. Mild strains of this virus have been reported from some countries. Magee (1948) described a member of Veimama plants in Australia, which showed very mild symptoms of BBTV infection. A mild strain of bunchy top which probably is widespread in Fiji and does little harm to the bananas can be used for cross-protection (Simmonds, 1982). Probably its presence alone has permitted Fiji's survival as a banana producing country for its impossible otherwise to explain the fact that the severe type of bunchy top is not nearly so common in the island as would be expected from its history.

As pointed out by Simmonds (1962) since the mild type of virus which is apparently not transmissible, banana suckers carrying it, would presumably have to be bulked up from suckers selected in diseased fields.

Using plants infected with mild strain of the virus a counter to "Bunchy top" in present cultivations or in areas where the rigid phytosanitary measures could not be adopted, deserves serious considerations. Natural mild strain protection had probably been the salvation of survival of banana industry in Fiji. Bhaktavatsalu *et al.* (1980) observed mild strain behaviour in hill banana var. 'Poovan'. Mild strain of the virus affecting Nendran (French plantain AAB) was reported by Nayar (1962) which though symptomless externally reduced fruit size and crop productivity.

1.21.6. Production of virus-free planting material

Wu and Su (1991) found that exposure of BBTV-infected plantlets of Cavendish banana to 40°C for 16 h daily for a period of upto 5 weeks did not free them from BBTV. However, BBTV was less readily transmitted by aphids from treated plantlets. When BBTV infected heat treated tissue was cultured for 3 months at 35°C some of the buds sprouted to produce healthy looking plantlets. The results suggest that uneven distribution and low concentration of virus after treatment at high temperature leads to BBTV-free primordial cells which in turn may develop into healthy plantlets.

1.21.7. Biological control

Aphid parasitoids are helpful in the control of aphid-borne virus diseases, since they reduce the population of potential vector aphids in crop plants. Two groups of parasitoids attack aphids. One group is the Aphididae and the other is *Aphelinus* and its allied genera of Aphilinidae. Parasitoids can be classified into two groups; natural parasitoids and potential parasitoids. The term natural parasitoids in this context means that the association with the aphid concerned has been confirmed in the field.

Parasitoids of *P. nigronervosa*

Lysiphlebus testaceipes has been ascertained to be a natural parasitoid of *P. nigronervosa* on banana in Cuba (Stary, 1966). This is the only record of a natural parasitoid of *P. nigronervosa. Aphidius colemani* may also be a natural parasitoid of the aphid. *A. colemani*, a native of Australia, was utilized for the biological control of *P. nigronervosa* in Tonga. Both *A. colemani* and *P. nigronervosa* are found in Australia, but the host-parasitoid association has not been reported in any scientific paper (Takada, 1992).

Potential parasitoids

Voelkl *et al.* (1990) tested five aphid parasitoids for their suitability as biological control agent of *P. nigronervosa*. These were *Lysiphlebus testaceipes* (a field strain from France and a Laboratory strain from Italy), *L. fabarum, Trioxys angelicae, Aphidius colemani* (a laboratory strain from East Africa) and *Ephedrus plagiator*. They concluded that of the five species tested, *L. testaceipes* and *A. colemani* were at least potentially suitable biological control agents against *P. nigronervosa*. Both species were successful mass reared on *P. nigronervosa* for five generations in the laboratory, and showed no obvious preference for certain host instars or morphs. When *E. cerasicola* originating from Norway was released in a glasshouse, it successfully propagated in *P. nigronervosa* on banana throughout the year.

Attempts to release *L. testaceipes* in Tonga

In 1986 and 1987, *L. testaceipes* was introduced into Tonga to control *P. nigronervosa*. Stechmann and Voelkl (1988) have reported on this attempt as biological control. A survey of the introduced parasitoid in Tonga found that fairly large colonies of *P. nigronervosa* were abundant on banana throughout the year (Stary and Stechmann, 1991).

Foraging patterns of *L. testaceipes* and *A. colemani*

To clarify the effect of forging behaviour, Stadler and Voelkl (1991) analyzed the foraging patterns of *L. testaceipes* and *A. colemani* in detail. They distinguished between exposed and concealed areas of the banana plant and found that small *P. nigronervosa* colonies fed mainly in concealed areas. With increasing colony size, colonies spread into exposed areas. The properties of aphids feeding in exposed parts of the plant was about 10% where small colonies were concerned, but about 90% of aphids in large colonies. *L. testaceipes* searched mainly in exposed areas, and made only short examinations of concealed areas. In contrast, *A. colemani* searched both exposed and concealed areas for hosts. These results suggest that aphid feeding in concealed areas would be attacked by *A. colemani* more often than by *L. testaceipes*.

Utilization of *A. colemani* for biological control

A. colemani is distributed in South America, East Africa, the Mediterranean, Central Asia, India and South Australia. Stary suggested that its original distribution area was northern India. From these it spread to the Mediterranean area and East Africa, and later to South America and South Australia. The host range of *A. colemani* is wide. Stary (1975) has listed 37 species of aphids as its hosts. It should be noted that this host range differs significantly according to region, and is generally much wider in the tropics than in the Mediterranean area or Central Asia.

If we are to introduce *A. colemani* for the biological control of *P. nigronervosa*, we should select the most effective strain. If we are to consider the Taiwanese form, we must confirm whether it accepts, *P.*

nigronervosa as a host or not. The form originating from East Africa has been confirmed to be a good biological control agent of *P. nigronervosa* by the suitability tests mentioned above. Concerning the South American form, no information is available about its association with *P. nigronervosa*. As stated above, *L. testaceipes* and possibly *A. colemani* are known to be natural parasitoids of *P. nigronervosa*.

Transgenic resistance

The two essential components from which to develop virus resistant transgenic bananas, a transformation and regeneration system and the nucleotide sequence of most if not all of the BBTV genome, are now available. Recently, both May *et al.* (1995) and Sagi *et al.* (1995, 2000) reported the regeneration of genetically transformed bananas. May *et al.* (1995) transformed the cultivar Grand Nain using Agrobacterium-mediated transformation while Sagi *et al.* (1995) transformed the cultivar Bluggose using microprojectile bombardment.

1.22. FUTURE LINE OF WORK

Dale (1987) in his review article on banana bunchy top, has given a comprehensive account of future strategy and possible directions on BBTV. The eradication campaign of BBTV in Australia had been successful in keeping the disease spread in check. The first proclamations and regulations under the New South Wales Plant Disease Act were officially announced in November, 1927 (Magee, 1936). For campaign to be effective, little more information about the virus was required and, indeed in the ensuring 73 years little more has been discovered. Most of the research has been confined to verification of the identity of the causal virus, varietal reaction, and epidemiology, etc.

Strategy for future research and control methods not only includes what could be achieved by using the latest technology available at present but also what may become available in the near future. Dale (1987) described "first-generation" and "Second-generation" directions for banana bunchy top control. The present or first generation measures depend upon a knowledge of the biological properties of the virus such as natural host range and mode of transmission etc. First generation control measures include eradication, supply of virus-free planting material, breeding or selection for immunity, resistance or tolerance; vector control; disease avoidance; and cross-protection.

The first most important aspect, which needs immediate attention, is to develop technique, which would have immediate application to quarantine, eradication and resistance breeding program. Summanwar and Marathe (1982) have recommended the use of 1% solution 2, 3, 5-triphenyl tetrazolium chloride for detecting infection of BBTV. However, this technique gives erratic results and cannot be recommended for use.

Jayaraj (1987) has recommended the use of Benedict solution, which gives brick red coagulation with juice from BBTV-infected plants whereas pink precipitation occurs with juice from healthy plants.

Based on present evidence, resistance of BBTV in bananas may have to be sought using the technique of somaclonal variation, mutation inducement and protoplast fusion to develop genetic variability. These techniques are already being used in Queensland and elsewhere to attempt to develop resistance to other banana diseases (Dale, 1987).

Control of aphid vector of BBTV seems to be expensive and would be successful for long-term control. Avoidance is more applicable to annual crops where planting times can be manipulated to avoid population peaks of aphid vectors.

Use of mild strain of BBTV may also provide encouraging results, however, this needs lot of effort and experimentation.

A sensitive assay system based on antiserum, like enzyme linked immunosorbent assay (ELISA) or immunosorbent electron microscopy (ISEM) or cloned cDNA probe would be of immense use on BBTV detection and control. Wu (1987) and Wu and Su (1990 a, b) have successfully purified BBTV and were able to produce monoclonal antibodies against this virus. Now the BBTV antiserum is commercially available.

Another approach, to get virus-free planting material would be, raising plants through tissue culture and detection of BBTV in these plants by serological and molecular techniques. There exists a misconcept that tissue culture raised are free from viruses. Tissue culture is a good technique for mass propagation of banana planting material but it does not necessarily eliminate the viruses. Many private companies in India are producing tissue culture banana plants and supplying to the farmers but these plantlets are not indexed for virus infection. This may lead to dissemination of viruses in the field and also in areas which were free from viruses. This practice must be stopped otherwise it is going to pose a real threat to banana cultivation. At least the mother suckers must be indexed for freedom from viruses. (Singh and Raj Verma, 2002; Drew *et al.*, 1989).

Suckers are more resistant to virus infection as compared to tissue cultured raised plantlets.

Greenhouse studies have shown that tolerance to BBTV and BSV is correlated with the size of banana plantlets.

The "second generation" control methods will, in all probability, utilize recombinant DNA technology such as gene transfer. This may well depend on the identification and isolation of BBTV resistance genes in wild *Musa* species or from outside the genus *Musa* or the family Musaceae. Success may depend on the manipulation of virus genes and therefore will require an intimate knowledge of the viral genome (Dale, 1987). Even though there is no recognized transformation system for the *Musa* (or for most Monocots), such systems are likely to be developed in the foreseeable future.

ANNEXURE -1

PROTOCOL FOR THE DETECTION OF BBTV BY TRIPLE ANTIBODY SANDWICH

ELISA (Thomas, 1991)

1. **Coating of Polystyrene Plate**
   - Dentate immulon 3 or Nunc Maxisorb plates used.
   - BBTV polyclonal Ig fraction, diluted in Carbonate coating buffer to 3 μg/ml.
   - Add 50 μl/well.
   - Incubate for 2-3 hours at room temperature (RT).
   - Wash 3 x 3 min with PBS-T.
2. **Antigen (Test Sample)**
   - Banana Midrib ground in a mortar and pestle with a pinch of acid washed sand and cold 0.2 M potassium phosphate buffer, pH 7.4 containing 0.5% sodium sulphite. Dilution 1 g/4 ml buffer.
   - Add 50 μl/well. Incubate overnight at 5°C.
   - Wash 3 x 3 min with PBS-T.
3. **Blocking**
   - 5% skim milk powder in PBS-T.
   - Add 200 μl/well.
   - Incubate 1/2 -1 hour at RT.
   - Wash 3 x 3 min with PBS-T.
4. **Tissue Culture Supernatants (Monoclonal Anti-Body)**
   - Diluted 1:1 to 1:10 with PBS-T-PVP, as appropriate for each antibody.
   - Add 50 μl/well,
   - Incubate at 5°C overnight or 2 hours at RT.
   - Wash 3 x 3 min with PBS-T.
5. **Conjugate**
   - Rabbit anti-mouse 19 G-Alkaline Phosphatase conjugate (ICN), diluted 1/2000 in PBS-T-PVP
   - Add 50 μl/ Well.
   - Incubate 3 hours at RT.
   - Wash 3 x 3 min with PBS-T.
6. **Substrate**
   - P-nitrophenyl phosphate (1 mg/ml) in diethanolamine buffer.
   - Add 100 μl/well.
   - Incubate 30 min to 2 hours at RT, then overnight at 5°C if necessary.
   - Read absorbance at 405 nm.

2

BANANA INFECTIOUS CHLOROSIS

Synonyms: Banana mosaic, Heart rot.

2.1. GEOGRAPHICAL DISTRIBUTION

Banana mosaic or infectious chlorosis is one of the most important and widely distributed virus diseases of banana. Its presence has been reported from various countries in the world, namely New South Wales, Australia (Magee, 1930); Brazil (Wardlaw and Mc Guire, 1933); Trinidad (Wardlaw, 1934): Davao (Calinisan, 1938); Haiti (Wardlaw, 1940); Borneo (Reinking, 1950); Cameroons (Wardlaw, 1948); Columbia (Ciferri, 1949); Somaliland (Castellani, 1955); Puerto Rico (Bird and Wellman, 1962); French Antilles and Ivory Coast (Yot-Dhauthy and Bove, 1966); Crete (Avgelis, 1987); India (Capoor and Varma, 1968; Joshi and Joshi; 1974); Surinam (Van-Hoof, 1962). Details of the geographical distribution have been presented in table 2.

In India, Banana mosaic was first reported widespread in Jalgaon district in Maharashtra state in 1943, but was attributed that time to lack of proper manuring and irrigation. Later on a sudden outbreak of a mosaic disease affecting the Dwarf Cavendish banana commercially grown in Eastern Khandesh, Surat, Thane, Nasik and Poona districts was observed in November, 1949 (Kamat and Patel, 1951; Desai and Bhide, 1956; Capoor, 1959; Desai, 1963). Now the disease is known to occur in almost all the banana growing states of the country. Desai (1963) reported the results of survey of 16,078 acres comprising 9047 gardens in Maharashtra state where he found 7661 banana plantations free from banana mosaic. In 1342 gardens the incidence was upto 0.5%; in 26 gardens from 0.5 to 1%; in 13 gardens from 1 to 2%; in 3 gardens from 2 to 3% and in 2 gardens above 5%. Mali and Deshpande (1976) surveyed the Marathwada region and reported the incidence upto 23% in Parbhani district of Maharashtra state. The incidence of banana mosaic may reach as high as 100% in some localities in Gujarat state. When plants are infected early, sheath rot and heart rot are commonly noticed. Joshi and Joshi (1974) stated that taking into account about 15,000 hectares under banana cultivation in Gujarat state and conservatively only 5% infection, the yield loss due to mosaic in South Gujarat was more than 10,000 tons, which amounts to more than 3 millions rupees every year. Rao (1980b) reported the incidence of banana mosaic disease in banana cultivars Robusta, Dwarf Cavendish and Rasthali from 2 to 7% in parts of Andhra Pradesh and Tamil Nadu. Kadhirvel *et al.* (1986) reported 20-80% incidence of banana mosaic in cv. Poovan from Trichi district in Tamil Nadu. The incidence was more in ratoon crops. Lassoudiere (1979) recorded 3.5% loss in the main weight of stem when incidence was 10%. Effect on yield due to dotted line mosaic was apparent in ratoon crop. Mosaic caused serious losses in Honduras in a farm on 648 acres, where more than 2000 cases were destroyed, or slightly more than 3 cases per acre during first year after planting (Stover, 1972). Thereafter disease incidence declined to an occasional case in several thousand acres. Mosaic is rare in old, well established plantations and is primarily a problem in localized areas only during the first year or two after planting in new land. When planting is immediately done after destroying established bananas, incidence is rare in contrast to planting after bush and weed fallow. The economic importance of mild 'Symptomless'strains of banana mosaic has not be determined.

Table 2. Geographical distribution of Infectious Chlorosis

Place	*Reference*
Australia (NSW)	Magee (1930); Burt (1988).
Borneo	Reinking (1950).
Brazil	Wardlaw and Mc Guire (1933), Zambolim *et al.* (1994), Barbosa *et al.* (2000); Pietersen *et al.* (1998); Trindade *et al.* (1998), Eiras *et al.* (2001).
Cameroons	Wardlaw (1948).
Central America	Waite (1960).
China	Kao and Faan (1983).
Columbia	Ciferri (1949); Casteno *et al.* (1994)
Crete	Avgelis (1987).
Davao	Calinisan (1938).
French Antilles	Yot-Dhauthy and Bove (1966); Lassoudiere (1979).
Egypt	Wardlaw (1972).
Ghana	Wardlaw (1940).
Haite	Osei *et al.* (1995).
India	Kamat and Patel (1951); Desai and Bhide (1956); Capoor (1959); Mali and Rajegore (1980); Estallita *et al.* (1996)
Indonesia	Feaken (1972).
Ivory Coast	Yot-Dhauthy and Bove (1966); Lassoudiere (1979).
Israel	Nitzany (1975).
Lebanon	Wardlaw (1972).
Morocco	Bouhida and Lockhart (1990).
Philippines	Sanjuan (1985); Magnaye and Eloja (1968).
Puerto Rico	Bird and Wellman (1962); Niblett *et al.* (1994).
Somali Island	Castellani (1955).
Surinam	Van Hoof (1962).
Tanzania	Vuylsteke *et al.* (1998).
Trinidad	Wardlaw (1934).
Taiwan	Hwang (1991).
Zanziber	Vuylsteke *et al.* (1998).
South Africa	Pietersen *et al.* (1998).

Mohan and Lakshmanan (1988) reported an outbreak of banana mosaic caused by a strain of cucumber mosaic virus (CMV), in all districts of Tamil Nadu on the cvs. 'Robusta' and 'Poovan'. CMV in banana has become more prevalent in Indonesia in recent years.

However, more information is needed on this disease in Indonesia. CMV is also a serious problem in Philippines (Magnaye and Eloja, 1968; Sanjuan, 1985). This is widespread among local cultivars such as the cooking banana 'Saba' and 'Cardaba' and the fancy variety 'Marado'. This has raised growing concern on the behaviour of the mosaic disease in Philippines.

CMV in banana was first observed in Central Taiwan in 1959, and have been commonly found in the south since 1970. It has become a major concern in recent years in banana plantations established with meristem-derived plantlets. Recent study has shown that these plantlets are more susceptible to CMV than the suckers, which are the conventional planting material.

Two strains of CMV namely the "Common" and "Leguminous" strains, have been isolated from diseased bananas. The common strain is more virulent and causes more prominent mosaic symptoms than the Leguminous strain. A cruciferae strain isolated from radish also attacks banana. The isolates causing mild mottle or chlorotic streak were identified to be legume strain (L-CMV). This isolate infects broad bean and cowpea inducing mosaic symptoms while tobacco plants show mild symptoms.

2.2. SYMPTOMATOLOGY

Symptoms vary greatly from mild streaking on the leaves to stunting, internal necrosis and death of the plant. Common or mild strain induce a diffuse mosaic (Fig. 6). Stunted plants have a rosette appearance due to bunching of leaves. Leaves may be stunted, narrow and strap shaped, and chlorotic (Fig. 7). The outer leaf sheath may separate from the pseudostem, and the emerging heart leaf may be necrotic (Fig. 8, 9). The most invariable symptom is the presence of yellow streaking on the other leaves, with the streaks usually running from the mid rib to the edge of the blade (Fig. 10). The streaks may be continuous or broken and in some cases may form elongated irregular bands with a ring spot type pattern (Fig. 11). Pockets of rot occur throughout the leaf sheath of the pseudostem in severe cases. In mild attacks the plant may recover and new leaves and suckers be symptomless. However, symptoms will again appear at varying intervals. When suckers from virus infected plants of different cultivars are planted in the glass house, there is no predictable pattern of symptom expression, but growth is slower in all plants from virus sources than in those from healthy plants. When plants are infected early, sheath rot and heart rot are commonly noticed. Such plants do not reach maturity or occasionally produce a bunch of fruits. The effect of secondary infection on plant growth and yield depends upon the age of the plants at the time of infection. Fruits from infected plants may appear normal, may be stunted with chlorotic streaking (Fig. 12), or may have a characteristic necrosis. Stems from virus-infected plants are generally smaller than those from healthy plants.

In heart-rot phase, the external manifestation of the disease is heart leaf necrosis, sheath rot down to suckers and concomittant death of plants. Occasionally chlorotic bands on older leaves are evident. Vertical and cross-sections of pseudostem reveal the downward extension of sheath rot to suckers and linear necrotic patches in pseudostem leaf bases. These symptoms are most prominent during winter. During summer, heart-rot phase is generally absent and symptoms on older leaves are only evident (Mali and Deshpande, 1976; Mali and Rajegore, 1980). Rao (1980a,b) reported a new strain of CMV from Andhra Pradesh. The symptoms include whip-like central top leaf, with older leaves below the terminal leaves for rolling off their margins. The leaf lamina size is reduced and the margins become wavy. Different types of mosaic associated with oak leaf-pattern with chlorotic granular bands spread all over the leaf (Fig. 13, 14). The leaf sheath area is sometimes more than that of leaf size and often give bunchy-top appearance of 'Filaree' type symptoms. The disease incidence has been found to be 2-7% in 'Robusta' or 'Dwarf Cavendish' varieties. Tissue raised plants are more susceptible to CMV than suckers.

2.3. TRANSMISSION

Virus from banana plants or suckers exhibiting the symptoms of mosaic mottling, chlorosis, or heart rot is readily transmitted

Fig. 6

Fig. 7

Fig. 8

Fig. 9

Fig. 10

Fig. 11

Fig.12

Fig.13

Fig.14

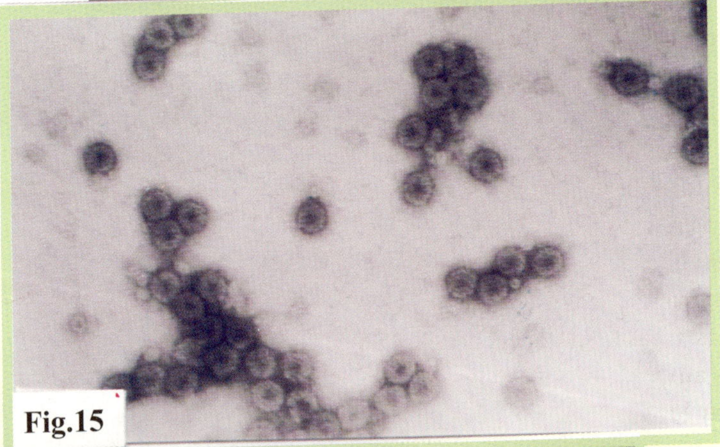
Fig.15

Fig.16

mechanically to a wide range of hosts but not to banana. Patil (1982) reported that CMV from banana could be transmitted by rhizome inoculation. Dipping of rhizome for 15 minutes in infected sap gives best results. Dipping and pricking method is also successful. Allam *et al.* (1995) reported soil transmission of banana mosaic virus (CMV) at a rate of 11.4%.

2.3.1. Aphid Vectors

Several aphids have been reported to transmit the disease in non-persistent manner. Transmission has been achieved with *Aphis gossypii, Rhopalosiphum maidis, R. prunifoliae, Myzus persicae, Macrosiphum pisi* and *Aphis craccivora* (Magee, 1940 a, Waite, 1960; Bird and Wellman, 1962; Van Hoof, 1962: Capoor and Varma, 1968; Evers, 1968; Mali and Rajegore, 1980). Some investigators (Magee, 1930; Calinisan, 1938; Silberschmidt and Nobrega, 1941) have reported the successful transmission of banana mosaic by the banana aphid, *P. nigronervosa* whereas others have reported negative transmission (Bird and Wellman, 1962; Mali and Rajegore, 1980; Rao 1980b). Most of these aphid transmission studies have been carried out from mosaic infected banana to other hosts and transmission tests from banana to banana were either not attempted or gave negative results. Transmission from banana to banana in the field is very low.

Bird and Wellman (1962) transmitted successfully the banana mosaic disease by *A. gossypii* and poorly by *A. craccivora* but not at all by *P. nigronervosa* in Puerto Rico. Van Hoof (1962) reported that in Surinam the banana mosaic virus was transmitted from infected to healthy bananas, to cucumber by *A. gossypii* and more rarely by *M. persicae* but not at all by *P. nigronervosa.*

Becquer and Bencomo (1974) reported from Cuba that aphid species *viz. R. maidis, Dactynotus ambrosiae* and *M. persicae* were found to transmit the virus (CMV) in banana plantations. Mali and Rajegore (1980) reported successful transmission of banana mosaic virus to test plants by *A. craccivora*. Rao (1980b) got successful transmission of banana mosaic by *A. gossypii* and *M. persicae.*

2.4. ETIOLOGY

There is great confusion about the virus (es) involved in inducing infectious chlorosis, heart rot or mosaic symptoms in banana. Most of the conclusions about the involvement of cucumber mosaic virus (CMV) in causing these symptoms have been derived from physical properties, host range and transmission studies. Some investigators in India considered banana mosaic and heart rot as separate diseases (Capoor and Varma, 1968) whereas others considered heart rot as a syndrome of banana mosaic (Joshi and Joshi, 1974; Mali and Rajegore, 1980). From the available literature in India, there seems to be three different strains of CMV inducing chlorosis or mosaic in banana (Capoor and Varma, 1970; Mali and Rajegore, 1980; Rao, 1980 b). Yot-Dhauthy and Bove (1966) purified the strain attacking banana and showed that it consisted of spherical virus particles similar to those of CMV. Undoubtedly several strains are present in banana. These cause symptoms varying from a mild yellow-streaking on the older leaves, with symptoms disappearing completely for various times, to severe necrosis and death of the plant. Allam *et al.* (1995) observed isometric particles in cytoplasm of infected banana cells, disorganized chloroplasts, degenerated mitochondria and inclusion bodies. Based on symptomatology, host range, physical properties, serology and electron microscopy, the virus causing banana mosaic has been identified as a strain of CMV and the member of Cucumovirus group (Gibbs and Harrison, 1970; Harrison *et al.*, 1971).

2.5. PHYSICAL PROPERTIES OF THE VIRUS

The virus is inactivated between 65-70°C and at a dilution of 10^{-4}. The virus remains viable upto 92 but not at 104 hours at room temperature (27-30°C) (Mali and Rajegore, 1980). Colariccio *et al.* (1996) reported dilution end point of 10^{-5}, thermal inactivation point between 70-75°C and longevity *in vitro* at room temperature for 24 hr. Virus infectivity was maintained at—20°C for 12 months in dehydrated leaf samples of *Musa* spp and *Nicotiana glutinosa*.

2.6. SEROLOGY

The virus is highly immunogenic. Antisera (polyclonal and monoclonal) are commercially available.

Mali and Rajegore (1980) found that the banana mosaic virus precipitated against the antiserum of CMV using the tube ring interphase and agargel diffusion techniques. ELISA and ISEM techniques have been commonly used for the detection of the virus.

2.7. PURIFICATION OF THE VIRUS

Yot-Dauthy and Bove (1966) and Colariccio *et al.* (1996) purified the virus involved in banana mosaic and studied its serology and other characters. The absorption spectrum showed that the virus is a nucleo-protein with a maximum at 259 to 260 nm and minimum 240 nm.

2.8. PARTICLE MORPHOLOGY

The virus particles are of polyhedric shape measuring about 26-28 nm in diameter (Fig. 15) and possesses a tripartite single-stranded RNA (ssRNA) genome.

2.9. HOST RANGE

The banana mosaic virus has got wide host range (Table 3). The cucurbitaceous, leguminous, solanaceous plants and weed species belong to *Solanum, Phytolacca, Physalis,* Canna, Commelina, *Musa, Datura,* Centaurea, and *Dianthus* genera have been reported to be important hosts in sporadic outbreaks of banana mosaic transmitted by aphid species in Australia, Honduras, Puerto Rico, Brazil, India and Surinam (Bird and Wellman, 1962; Evers, 1968; Magee, 1940b; Silberschmidt and Nobrega, 1941; Van Hoof, 1962; Waite,1960; Mali and Rajegore, 1980; Avgelis, 1987).

In India, Capoor and Varma (1970) reported only cucumber as the additional host of banana mosaic whereas Mali and Rajegore

(1980), Rao (1980b) and Joshi & Joshi (1976) reported a wide host range (Table 3).

Broad leaf weed flora in banana plantations which contain many species carrying CMV such as *Physalis* sp., *Isotoma longiflora* (a symptomless host), *Ricinus communis* and species of *Cucumis, Cucurbita, Crotalaria* and *Pucraria* have been reported by Waite (1960) from Central America. *Commelina* (Fig. 16) in banana plantations has been reported to be often infected with CMV that can be transmitted to bananas (Bird and Wellman, 1962; Niblett *et al.*, 1994). Salih and Nas (1974) reported several species of weeds *viz.*, *Carastium arvensa, Eupatorium dubium, Rorippa islandica, Solanum dulcamaras* and *Verbasculum thapsus* to be hosts of virus causing banana mosaic.

Joshi and Joshi (1976) reported cucumber, *Phaseolus vulgaris, Vigna sinensis, Datura stramonium, Zinnia elegans, Chrysanthemum, Centraurea cyamus* and *Dianthus barbatus* to be the host of mosaic of banana. Mali and Rajegore (1980) reported several weeds to be the hosts of banana mosaic virus. *e.g. Achyranthus, Amaranthus, Celosia, Viscosa, Datura, Physalis* and *Solanum* species. Patil (1982) reported some new hosts for banana mosaic virus (CMV) *viz. Amarathus spinosa, Digera arvenis, Canavalia ensiformis, Cyamopsis tetragonoloba, Datura fastuosa, Nicotiana megalosiphon* and *Corohorus acutangulus*.

There seems to be some variations in the host range and the reaction of different plant species to banana mosaic virus, as reported by different workers (Mali and Rajegore, 1980; Avgelis, 1987). The host range mentioned by these authors have been presented in tables 3 and 4. These variation may be attributed to the variations in the strains of CMV in different places. Also the cultivars used may play a role in the variation in the symptoms. CMV has an extensive host range with nearly 800 species of plant recorded as host (Douine *et al.*, 1979) and is transmitted by over 60 species of aphids (Francki *et al.*, 1979).

Table 3. Host range and reaction of cucumber mosaic virus (CMV) (Mali and Rajegore, 1980)

Hosts	*Reaction*	
	Local	*Systemic*
Amaranthaceae		
Achyranthes asper	NLL	Sev, M
Amaranthus polygamus	CLL	M, Mt
Celosia argentea	RLL	R, Mt, Ld
Gomphrena globosa	NLL	—
Viscosa angulata	—	Mm
Chenopodiaceae		
Chenopodium album	NLL	—
C. amaranticolor	NLL	—
C. murale	NLL	—
C. quinoa	NLL	—
Commelinaceae		
Commelina benghalensis	NLL	Latent
Compositae		
Bidens pilosa	—	—
Carthamus tinctorius	—	M
Helianthus annuus cv. EC 68414	—	M
Sonchus arvensis	—	—
Tridex procumbens	NLL	—
Xanthium strumarium	—	Mt, Ld
Zinnia elegans	—	Bl, M
Cucurbitaceae		
Benincasa hispida	—	M, Ld
Citrullus vulgaris cv. Sugar Baby	—	M
Cucumis melo	—	—
C melo var. utilissimus	—	Vc, Mt
Cucurbita moschata	—	Intv. Chl
C. pepo cv. Petti Pan	—	Intv. Chl
Cucumis sativus cv. Bangalore Dwarf	—	Vc, M, Ld
Lagenaria siceraria	—	—
Luffa acutangula	—	Vc, Mt

L. cylindrica	—	—
Momordica charantia	—	—
Euphorbiaceae		
Euphorbia geniculata	—	Mm
Graminaceae		
Zea mays cv. Double Deccan hybrid	—	—
Labiatae		
Ocimum basilicum	—	—
O. sanctum	—	—
Leguminosae		
Arachis hypogaea cv. SB Xl	—	—
Canavalia ensiformis	—	—
Cassia tora	—	—
Cicer arietinum cv. H-208	—	—
Crotalaria juncea	—	—
Cyamposis tetragonoloba cv. Pusa Navbahar	—	—
Dolichos biflorus	—	—
Glycine max cv. Bragg	—	—
Phaseolus aconitifolius	—	—
P. aureus cv. Pusa Vaishaki	—	—
P.lunatus	—	Vc, Mt, Ld
P. mungo cv. Sindkheda	NLL	—
P. vulgaris cv. Prince	NLL	—
P. vulgaris cv. Master piece	—	—
Vicia faba cv. Ec-5063	—	—
Vigna unguiculata cv. Black-eye	NLL	M
- - cv. P-1473	NLL	M, Mt
- - cv. Early Ramshorn	NLL	—
- - cv. Shirpur Local	NLL	Mt
Malvaceae		
Abelmoschus esculentus cv. Pusa Sawani	—	—
Abutilon indicum	—	—

Solanaceae		
Capsicum annuum cv. NP-46A	—	M, Mt
Datura metel	—	Lp, Mt
D stramonium	CLL	M
Lycopersicon esculentum cv. Pusa Ruby	—	Vc, M, Ln
Nicotiana glutinosa	—	Bl, M, Ld
N. tabacum cv. White Burley	—	M, Mt
cv. Xanthi	RLL	R, Lp
Petunia axillaris	—	M, Bl, Ld
P. violacea	—	M, Bl, Ld
Solanum melongena cv. American Black Beauty	—	R, Olp
S. khasianum	—	M, Ld
S. nigrum	—	Gr, Vb, Mt
Tiliaceae		
Corchorus acutangulus	—	—
Sesamum indicum	—	—

*Bl = Blistering; Chl = Chlorosis; CLL = Chlorotic local lesions; Gr = Green; Intv = Inter-veinal; Ld = Leaf distortion; Le = Leaf elongation; Ln = Leaf narrowing; Lp = Line pattern; M= Mosaic; Mm = Mild mosaic; Mt = Mottle; NLL = Necrotic local lesions; Olp = Oak-leaf pattern; RLL = Ring-spot local lesions; R = Rings; Vb = Vein banding; Vc = Vein clearing.

Table 4. Host range of Cucumber mosaic virus isolated from Banana (Avgelis, 1987)

Hosts	*Reaction**	
	Local	*Systemic*
Amaranthaceae		
Celosia cristata	—	Mt
Gomphrena globosa	ReNL	M
Chenopodiaceae		
Chenopodium amaranticolor	NL	—
C. quinoa	NL	—
Spinacia oleracia	—	M

Compositae		
Helianthus annuus	NL	—
Lactuca sativa	—	M
Zinnia elegans	—	M
Cucurbitaceae		
Benincasa cerifera	NL	Mt
Citrullus vulgaris	NL	—
Cucumis melo	CL	M
C. sativus	CL	M
Cucurbita ficifolia	NL	—
C. foetidissima	—	—
C. maxima	—	—
C. pepo	C-NL	M
Lagenaria vulgaris	—	M
Luffa acutangula	RN	VC
L. cylindrica	NL	VC, Mt
Momordica balsamina	NL	Mt
Labiatae		
Ocimum basilicum	—	—
Leguminosae		
Phaseolus aureus	ReNL	(L)
P. vulgaris cv. 'Pinto'	—	(L)
cv. 'La Victoire,	—	(L)
Pisum sativum	ReNL	(L)
Vicia faba	ReNL	(L)
Vigna sinensis	ReNL	(L)
Solanaceae		
Capsicum annuum	—	M
Datura metel	—	M
D. stramonium	—	M
Lycopersicon esculentum	—	VC, D
Nicotiana benthamiana	—	VC, D
N. clevelandii	—	Mt, D
N. glutinosa	—	VC, M, D
N. megalosiphon	—	VC, N

N. paniculata	—	VC, M
N. rustica	—	VC, M
N. tabacum cv. 'Samsun'	—	VC, M
cv. 'WB ml'	—	VC, M
cv. 'WB nl'	—	VC, M
cv. 'Xanthi nc'	—	M
Petunia hybrida	—	VC, N

*CL = Chlorotic lesions, D = Deformation, M = Mosaic, Mt = Mottle, NL = Necrotic lesions, Re = Reddish, VC = Vein clearing, N = Top necrosis, (L) = No infection detected by back inoculation to *Vigna sinensis*, - = No symptoms.

2.10. VARIETAL SUSCEPTIBILITY

In *Musa*, cultivars of AAA and AAb are most commonly infected. Bluggoe, ABB is occasionally infected as well as diploids (AA, AB). In Central America, cucumber mosaic incidence has been for greater, in the Cavendish varieties than in Gross Michel (Stover, 1972). *Musa balbisiana* is the only *Musa* species in which symptoms of the virus infection have not been found in the field. However, *M. balbisiana* seedlings from Central America have been observed with whitish chlorotic mottling characteristic of virus infection even though virus symptoms are not seen on mature plants (Stover, 1972). In India Ela vazahiAthiya kel, Kare Bale cultivars and *M. balbisiana*, *M. chilicarpa*, *M. coccinia* and *M. accuminata* have been reported resistant. The Basrai and Lal Velchi banana are highly susceptible.

Capoor and Varma (1970) reported the following varieties susceptible to mosaic—CHAKKARAKELI: Thella Chakkarakeli, Rasthali, Rajeli or Raja Keli; VAMANAKELI (-) Basrai, Harichall, Mauritius, Cavendish, Pedda Pacha ARATTI, NENDRAN: Desi, Nendran, Singapuri; CHENKADALI (-) Betta Bale, Lal Velchi, Lal Kel, Vella Vazhai; RASTHALI (-) Amrithapani, Son Kel; NEY POOVAN (-) Kadali, Safed Velchi or Lady's Finger; POOVAN-Yarchi Bale, Metheli or Motta Poovan; NENDRA PADATH THI; VANNAN; PEYLADDAN-Peyan; NEY MANNAN (-) China Monthan; Monthan—Maduraga Bale, Nir bale or Nir Bontha; MISCELLANEOUS (-) Alshi, Amroli, Belhe, Bhur Kel, Bargi Bale,

Hazari, Harnai Khasadi, Kagdi Velchi, Kangal, Lokhandi, Mumbai Bale, Mysore Mitka, Mittli, Shendurni, Surat, Valha Sakas.

San Juan (1985) reported that local cultivars such as the cooking bananas, Saba and Cardoba, and the fancy varieties, 'Morado' are highly susceptible to banana mosaic.

2.11. STRAINS

Cucumber mosaic cucumovirus (CMV) is a multi components virus with a single stranded positive sense RNA genome consisting of three RNA species plus a subgenomic RNA (Palukaitis *et al.*, 1992). Numerous isolates differing in host range and pathogenicity have been reported. They fit into two major serotypes, subgroups I and II. Subgroup I, typified by isolate DTL occurs predominantly in the tropics and subtropics, while Subgroup II typified by isolate TORS, is prevalent in temperate regions (Hasse *et al.*, 1989).

Strains of CMV are characterized into subgroups on the basis of serological relationships, host-range, peptide mapping of the viral coat protein (CP) and nucleic acid hybridization (Edwards and Gonsalves, 1983). Hu *et al.* (1995) have cloned and sequenced the nucleotide and aminoacids of CMV of banana from Hawaii. Based on sequence homology, CMV in banana belongs to CMV subgroup I and shares 99% sequence identity with CMV strain. Another isolate (CMV-Oahu) shares 91% to 93% nucleotide and amino acid sequence identity respectively, with both CMV-C and CMV- Hawaii. CMV-Oahu isolate is less closely related to other strains in subgroup I than is the CMV - Hawaii isolate. Pappu *et al.* (1991) sequenced CMV strains in banana from Puerto Rico and found that all the strains belong to CMV serogroup I. Thomas (1991), however, has detected both CMV subgroups infecting banana in Australia. Zora Singh *et al.* (1995) isolated and identified CMV subgroup I from banana showing infectious chlorosis symptoms from Western Australia, using RT-PCR, restriction enzyme analysis and comparison of nucleotide sequences. Nucleotide sequencing showed around 98% homology to sequences of CMV subgroup I isolates but only 76 to homology of CMV- subgroup II isolates.

Yot-Dhauthy and Bove (1966) concluded that a number of strains of CMV are probably involved in the mosaic disease of *Musa* spp. These differ in virulence causing streaks, spots to more damaging types causing broader strands. Mali and Rajegore (1980) have reported the involvement of Y-strain of CMV in band type mosaic symptoms on banana leaves. In Taiwan, three strains of CMV namely the common, leguminous and cruciferous strains have been isolated from diseased bananas (Hwang, 1991).

2.12. DETECTION

The presence of the virus in the plants can be detected by the following methods:

(*a*) Symptomatology

The most conspicuous symptom is the presence of yellow streaking on the older leaves, with streaks usually running from the midrib to the edge of the blade. The streaks may be continuous or elongated. In heart rot phase, the external manifestation of the disease is heart leaf necrosis, sheath rot down to suckers and concomitant death of plants.

(*b*) Diagnostic hosts

The presence of CMV can be detected by inoculating indicator hosts, which produce necrotic local lesions (Tables 3 and 4).

(*c*) Serology

Serology is the most reliable and quick method of detection of CMV infection in banana (Thomas, 1991; Niblett *et al.*, 1994). These include, agar gel immuno diffusion, and microprecipitin tests. ELISA, is most reliable serological technique for the detection of CMV in banana plants and aphid vectors. There are two commercial ELISA tests available for CMV, namely, AGDIA and Sanofi. The antisera developed in Australia have not yet been compared with the commercial tests. It is believed that AGDIA will not detect all isolates of CMV. The protocol for the detection of CMV by Biotin/ Streptovidin ELISA has been given by Thomas (1991). CMV isolates can be grouped into three serotypes, termed TORS, DTL and Co. The first two groups are fairly closely related (Serological differentiation index; SDI: 1-2) while the Co. serotype is more distantly related (SDI:3) to both other groups. The division of isolates between the two major serotypes TORS and DTL is supported by nucleic acid hybridization studies (Owen and Palukaitis, 1988 and references therein). These serological differences have implications for the detection of field isolates of CMV by ELISA. It is widely recognized that coupling of antibodies with enzymes (*e.g.* alkaline phosphatase) decreases their affinity for their target antigens. This has minimal effect on the reaction with closely related antigens, but can seriously impedes the reaction with more distantly related ones (Thomas, 1991). With CMV, closely related virus isolates (SDI: 1-2) can be detected by DAS ELISA but more distantly related virus isolates (SDA > 3) may be undetectable (Devergne *et al.*, 1981). These problems can be circumvented by various forms of indirect ELISA. The common feature of these assays is that trapped virus particles are detected by native antibodies (*e.g.* TAS ELISA) or with antibodies conjugated in such a way that their binding efficiency is not impaired (*e.g.* biotinylation). Niblelt *et al.* (1994) classified banana isolates of CMV in their serotypes *i.e.*, CMV-Cr and CMV-VI (AGDIA, El.Khart, IN 46514). Immunosorbent electron microscopy (ISEM) is another very good test for the detection of CMV.

(*d*) Electron microscopy

Purified virus preparations can be seen in the transmission electron microscope for the presence of virus particles.

(*e*) RT-PCR detection

Presence of CMV in banana can easily be detected by RT-PCR technique (Zora Singh *et al.*, 1995). Restriction enzyme analysis of PCR products using Eco R1 and MSPI showed that the samples tested from diseased banana were infected with CMV subgroup I isolates. This result was confirmed by sequencing PCR-amplified products, which showed 98% homology to sequence of CMV subgroup I, but only 76% homology to subgroup II isolates. The

RT-PCR is a reliable assay to detect CMV in banana plants and, when combined with restriction enzyme digestion, provides a means of identifying the CMV subgroup present. The virus can also be detected by cDNA probe to RNA genome of CMV (Kiranmai *et al.*, 1997).

2.13. ULTRASTRUCTURAL CHANGES

El Deeb *et al.* (1997) studied the ultrastructural changes in banana plant infected with CMV and found that mesophyll cells of young leaves showed several ultrastructural changes due to virus infection. The cytoplasm contained few vacuoles and many vesicles. The plasmalemma was separated from the cell wall. Virus aggregates were detected in the cytoplasm, proplastids and nuclei. The chloroplasts of CMV infected cells showed swollen granal thylakoids and, in some cases, they appeared swollen and degenerated. In other cases, chloroplasts contained vesicles and osmophilic granules and an abundant number of mitochondria appeared in CMV infected cells. Elongated nuclei with virus-like particles were detectable cell-wall thickening projected irregularly due to CMV-infection.

2.14. MANAGEMENT

2.14.1. Legislation and spraying

Following two localized outbreaks of cucumber mosaic in Gross Michel in Honduras in 1955 and 1957, diseased plants and extensive areas surrounding healthy plants were sprayed with insecticides. Then the diseased plants were cut and sprayed with diesel oil. Adam (1962) described a reduction in virus incidence on Giant Cavendish from 21.2 in 1956 to 0.1 in 1960 in cases of 1000 acres in Honduras. Infected plants were removed. Twenty five acres around diseased plants were sprayed with nicotine sulphate at 20 and 40 day intervals following eradication. It has since been found that these expensive measures are not necessary and roguing of diseased plants alone has provided good control. After the rogued plants have decayed the area is re-planted. The mats are destroyed with herbicides. In India, to control and eradicate this disease, a scheme was sponsored by I.C.A.R., in 1957 (Desai, 1963). The disease was brought under check by destroying the diseased stools and planting disease-free suckers in new areas. The susceptible weed flora and inter cultivation of cucurbits with banana also play an important role in outbreaks of banana mosaic. Under these situations, weeds should be checked by spraying herbicides and inter cultivation of cucurbits should be stopped.

2.14.2. Meristem tip culture

Heat therapy of suckers and culturing of meristem tips alone is not effective in producing a high percentage of virus-free plants (Berg and Bustamante, 1974). However, when infected suckers were heat treated at 35-43°C and meristem from lateral buds, which developed 100 days after treatment, were grown on a modified Knudson's medium enriched with NAA, virus could not be detected in the plants, but the overall success was 75%. Gupta (1986) reported complete eradication of mosaic disease by heat therapy and meristem culture in various cultivars of banana (*M. acuminata* 'AAA' cvs. Grande Naine and Valary), and plantain *M. acuminata* × *M. balbisiana* 'ABB' cvs. Maricongo common dwarf and Super plantain). Suckers were subjected to heat therapy at 38-40°C for 14 days prior to the culture of their meristem tips. Plants derived from heat treated meristems of infected plants were free from the disease, as determined by visual inspection, mechanical inoculation to *Cucumis sativus* and electron microscopy (Gupta, 1986). Bondok *et al.* (1987) obtained CMV-free banana plants by shoot tip (0.5-1 mm) culture cultured on Murashige and Skoog mineral salt medium under or without antiviral compound (Thiouracil). Virus-free plants of banana cvs. Poyo and Giant Cavendish infected by CMV (Cucumovirus) were produced by meristem tissue culture. Explants were cultured in medium containing Murashige and Skoog mineral salts supplemented with Thiamine HCI (1 mg/lit), Pyrodoxine (1 mg/lit), Nicotinic acid (100 mg/lit) and Myo-inositol (100 mg/lit) and solidified with 7g/lit bacto agar. Virus indexing of plants was done by inoculating cowpea cv. Black eye as test plants. The efficiency of virus elimination was 68.8% for Poyo and 66.8% for Giant Cavendish (Rivas, 1988).

3

BANANA STREAK

3.1. GEOGRAPHICAL DISTRIBUTION

Viral leaf streak was first described as 'la mosaique a tirets' occurring on the banana cultivar Poyo in Nieky in Valley Cote d' Ivoire (Ivory Coast), Africa in 1974 (Lassoudiere, 1974; Dabek and Waller, 1990), following previous reports of a mosaic disease in Africa thought to be caused by a strain of cucumber mosaic virus (CMV). Despite the fact that these first recordings of Banana Streak Virus (BSV) were from Africa, in a relatively short period of time, the disease had been observed or identified in most banana and plantain producing areas of the world. This suggests that the causal agent, banana streak virus, has been widely distributed for many years, and that its geographical origin is still uncertain. Indeed photograph of what is almost certainly BSV infection have even appeared labeled as CMV infection in reference texts. The causal agent BSV was first identified in 1986 (Lockhart, 1986). The virus has since been identified in many of the *Musa* growing countries in the Americas, Africa, and Asia and is the most frequently occurring virus disease of *Musa*. However, although banana streak was described only relatively recently, there is good evidence that it is not a new disease. It has been undoubtedly frequently misdiagnosed as CMV infection (Lockhart, 1996), and illustrations of symptoms variously ascribed to CMV, mosaic, infectious chlorosis, and even physiological disorder (Yot-Dhauthy and Bove, 1966; Stover, 1972, p. 127; Wardlaw, 1972, pp. 116-145) are unmistakably those now known to be associated with BSV infection (Lockhart, 1986; Lockhart,1994a,1995; Dahal *et al.*, 1998a).

Current interest in BSV and the disease it causes has been sparked by two separate events. The first was the development of diagnostic methods (Lockhart and Autrey, 1988; Lockhart and Olszewski, 1993; Lockhart, 1996; Thottapilly *et al.*, 1997, 1998; Ndowora and Lockhart, 1998; Lockhart *et al.*, 1988) which made it possible to distinguish clearly between BSV and CMV infections, which can sometimes be easily confused. This led to an increasing number of reports of BSV occurrence, and the realization that the virus was of worldwide distribution. The second event was the observation that a significant proportion of improved *Musa* tetraploid hybrids, bred for disease and pest resistance and higher yield, were found to be infected by BSV under circumstances which appeared to preclude natural infection in the field. Geographical distribution of BSV has been given in table 5.

Table 5. Geographical distribution of banana streak disease

Region	*Country location**	*References*
Europe	Spain (Canary Islands)	Diekmann and Putter,1996
	Portugal (Madeira)	Jones and Lockhart,1993
Africa	Benin	Diekmann and Putter,1996
	Cameroon	Gauhl and Pasberg-Gauhl, 1997; Gauhl *et al.*, 1999 a, b
	Cape Verde	Diekmann and Putter, 1996
	Cote d' Ivoire	Yot-Dhauthy and Bove, 1966
	Ghana	Diekmann and Putter, 1996
	Gunea	Diekmann and Putter, 1996
	Kenya	Diekmann and Putter, 1996
	Madagascar	Jones and Lockhart, 1993
	Malawi	Vuylsteke *et al.*, 1996
	Mauritius	Jones and Lockhart, 1993
	Morocco	Lockhart, 1986
	Nigeria	Jones and Lockhart, 1993
	Rwanda	Sebasigari and Stover, 1988
	South Africa	Jones and Lockhart, 1993
	Tanzania (incl. Zanzibar)	Sebasigari and Stover, 1988; Vuylsteke *et al.*, 1998; Dabek and Waller, 1990.

	Uganda	Dabek and Waller, 1990; Tushmereirwe *et al.*, 1996; Kubiriba *et al.*, 1997
Latin America	Brazil	Jones and Lockhart, 1993
Canbbean	Colombia	Reichal *et al.*, 1997
	Costa Rica	Diekmann and Putter, 1996; Brunt *et al.*, 1987
	Cuba	Jones and Lockhart, 1993
	Ecuador	Jones and Lockhart, 1993
	Grenada	Jones and Lockhart, 1993
	Guadeloupe	Jones and Lockhart, 1993
	Haiti	Lockhart, 1999
	Honduras	Jones and Lockhart, 1993
	Jamaica	Jones and Lockhart, 1993
	Nicaragua	B.EL. Lockhart,1999 (personal communication)
	Puerto Rico	B. EL, Lockhart,1999 (personal communication)
	St. Lucia	B.EL.Lockhart,1999 (personal communication)
	Trinidad	Jones and Lockhart, 1993
	USA (Florida, Virgin Islands)	Diekmann and Putter, 1996
	Venezuela	Jones, 1995
Asia-Pacific	Australia	Thomas *et al.*, 1994
	China	Jones and Lockhart, 1993
	India	Singh *et al.*, 1996
	Indonesia	Diekmann and Putter, 1996
	Malaysia	Diekmann and Putter, 1996
	New Caledonia	Diekmann and Putter, 1996
	Papua New Guinea	Diekmann and Putter, 1996
	Philippines	Diekmann and Putter, 1996
	Sri Lanka	Diekmann and Putter, 1996
	Taiwan	Su *et al.*, 1997
	Thailand	Diekmann and Putter, 1996
	Tonga	Thomas *et al.*, 1994
	Vietnam	Diekmann and Putter, 1996
	Western Samoa	Thomas *et al.*, 1994

*Confirmed by electron microscopy or serology.

In India, the presence of BSV has been confirmed from States of Tamil Nadu, Kerala, Karnataka, Andhra Pradesh, West Bengal, Bihar, Assam, Maharashtra and Gujarat (Singh *et al.*, 1996; Selvarajan *et al.*, 1997; Ram and Singh, 1999). In Tamil Nadu, 60-70% plants had characteristic symptoms of BSV, which was more in 2nd and 3rd cycle of crop. In Maharashtra 5-10% incidence of BSV was noticed in *in vitro* propagated plants of Cavendish banana (Ram and Singh, 1999).

3.2. SYMPTOMATOLOGY

Naturally occurring isolates of BSV produce widely differing symptoms in the *Musa* genotypes in which they are found. Dahal *et al.* (1998) reported that the expression and distribution of symptoms was erratic within individual leaves as well as between different leaves of the same plant. These symptoms vary from inconspicuous chlorotic flecking to lethal systemic necrosis, and include yellow, brown and black streaking, Cigar-leaf necrosis, based pseudostem splitting, internal pseudostem necrosis, and formation of small, deformed bunches. "Traveller's palm" appearance of emerging leaves (*i.e.* leaves unfurling in a single plane) can be observed.

The symptoms of banana streak consist of broken or continuous chlorotic streaks (Fig. 17) and spindle shaped lesions that may be sparse or concentrated in distribution (Fig. 18). These chlorotic lesions become progressively necrotic and blackened and produce a characteristic black-streaked appearance in older leaves (Fig. 19). Symptoms may vary from faint broken chlorotic lines to necrosis of emerging leaves, internal necrosis of pseudostem and plant death. A conspicuous feature of this disease is the erratic distribution of symptoms, over individual leaves as well as on different leaves of the same plant. It is interesting that symptoms of heart-rot and internal pseudostem necrosis (Fig. 20, 21) were already described several decades ago (See Wardlaw, 1972, p. 123) to banana mosaic disease, again emphasizing the point that BSV infection in *Musa* is not of recent origin. Symptoms typically vary greatly during the course of the year. One or two leaves may emerge with pronounced symptoms, followed by a succession of leaves with inconspicuous or no symptoms, but which contain detectable levels of virus particles. A few months later, another flush of symptoms may appear. This periodicity of symptom expression is closely corrected with seasonal temperature fluctuations (Lockhart, 1996b; Dahal *et al.*, 1998b) and symptoms are more likely to be observed in plants growing in areas with marked seasonal temperature variations than in areas where temperature remain constant throughout the growing season.

Other symptoms associated with BSV include leaf bases falling away from the pseudostem (Fig. 22), narrow black streaks in the pseudostem (Fig. 23), thicker leaves cigar leaf necrosis and construction of the bunch on emergence, aberrant bunch emergence; reduced bunch size and distortion of fingers (Lassoudiere, 1979; Gauhl and Pasberg Gauhl, 1994). Peel splitting and necrotic streak and spot symptoms have been observed on fingers on 'Grand Nain' (AAA, Cavendish Subgroup) in Costa Rica (Fig. 24, 25) and Ecuador. In Australia, broad yellow lines in the leaf lamina parallel to the midrib, a purple margin to the leaf lamina, leaf twisting, grooves in the base of the pseudostem have also been observed in 'Williams' (AAA, Cavendish Subgroup) (Daniells *et al.*, 1998). Three phases of symptoms expression have been recognized in commercial Cavendish plantations in Ecuador. The first is the appearance of chlorotic streaks in the leaves (Fig. 26), the second is the appearance of dark blotches on the pseudostem and midrib and third is the splitting of the outer leaf sheaths of the pseudostem, sometimes as far up the plant as the petiole which allows the entry of bacteria that causes a soft rot of the base of the pseudostem.

In general, leaves initiated or produced during period of high temperature develop more pronounced symptoms. The symptoms of banana streak superficially resemble those caused by CMV infection but with practice, the two diseases can be readily distinguished. One characteristic feature of CMV infection is a midrib mosaic that is absent in leaves from plants infected with banana streak only. The isolates of CMV occurring in banana in Morocco is readily sap transmitted from banana to herbaceous test plants and occurring banana plantations, in weed hosts such as *Bryonia dioca* Jacq. and *Malva* sp.

3.2.1. Host factor affecting transmission and symptom expression

Different cultivars of banana show different degrees of disease resistance. The Peichiao of Taiwan Cavendish (AAA) produce severe typical streak symptoms, while only mild streak symptoms develop on Mysore (AAB) banana plants. No typical streak symptoms develop on Taiwanese and exotic cultivars of Latundan Banana (AAB). The infected Latundan plants are symptomless or produce mild chlorotic mottling (Su, 1998). The highest content of BSV was detected in the Cavendish plants, with lower content in Mysore plants and the lowest contents in plants of Latundan.

Su (1998) found that maturity of banana plants affected symptom expression. The smaller (less than 20 cm) tissue culture plantlets, inoculated produced severe streak symptoms. Some of these diseased plants developed severe leaf necrosis causing heart rot of spindle leaves characteristic of symptoms caused by CMV. The symptoms of young and spindle leaves were distinct and severe and became milder or indistinct towards maturity of the leaf.

Temperature can affect symptom development in Cavendish plants (Dahal *et al.*, 1998). Leaves which emerge during the cold winter (December to February) do not produce symptoms or produce very mild symptoms, while leaves emerging during spring (March to May) produce distinct symptoms. Milder symptoms develop or masked on leaves growing during the hot summer (June to August).

3.3. EFFECT OF THE VIRUS

The effect of BSV infection on plant growth, bunch yield and fruit quality may be very variable, for two reasons. Firstly, isolates of BSV differ greatly in the severity of symptoms produced, ranging from occasional chlorotic streaking to lethal systemic necrosis. Secondly, the symptoms caused by BSV infection characteristically occur sporadically over the course of the year and are correlated with

Fig.17

Fig.18

Fig.19

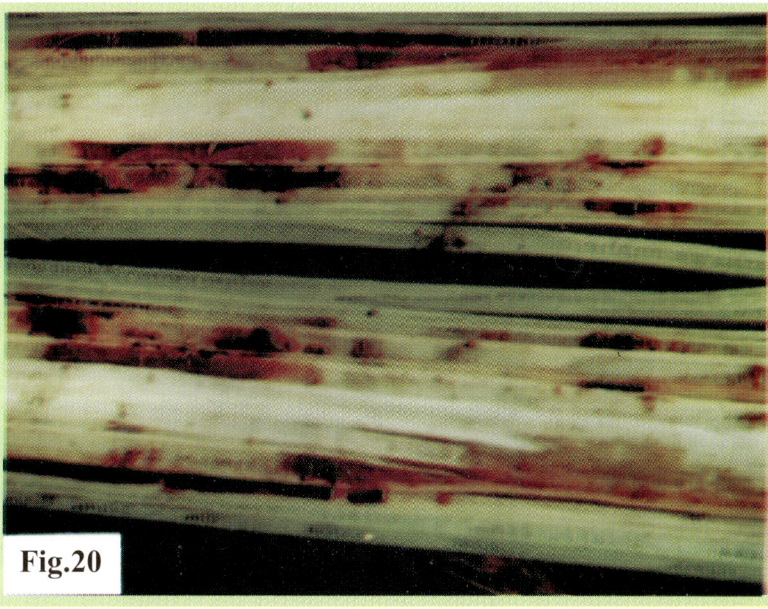

Fig.20

Fig.21

Fig.22

Fig. 23

Fig. 24

Fig. 25

Fig. 26

Fig. 27

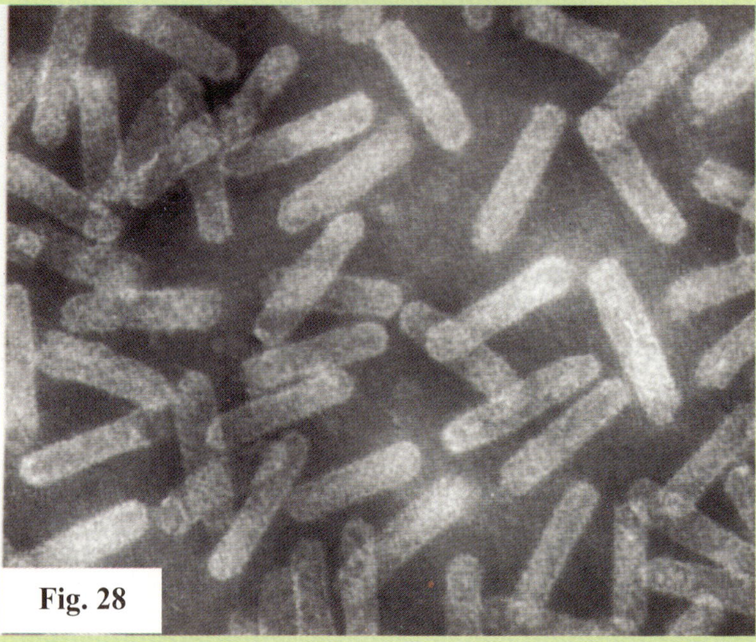

Fig. 28

variations in virus concentration. Yield losses due to BSV infection in banana have been documented for the AAA cultivar Poyo in Ivory Coast (Lassoudiere, 1979). Reduction in bunch size and malformation of fingers have also been reported for BSV infection in 'Dwarf Cavendish' (Lockhart, 1986). It has been suggested that BSV infection may have a significant effect on bunch size and fruit quality when floral initiation and early bunch development coincide with a period of increased virus synthesis (Lockhart and Olszewski, 1993). Plant death due to BSV has been recorded from Rwanda (Lockhart, unpublished) and Nigeria (Gauhl and Pasburg-Gauhl, unpublished). In variety Poovan yield loss of 48.8% has been recorded in Tamil Nadu (India) (Selvarajan *et al.*, 1997). Daniells *et al.* (2001) reported that BSV-CaV infection resulted in 18 day delay in harvest causing 6% reduction in yield per annuum.

Thomas *et al.* (1998) reported that the percentage of plants with high virus titres also increased during both cropping cycles, upto the time of bunch emergence, after which the percentage dropped again. The detectability and symptom expression of the virus in individual plants throughout the cropping cycle was erratic.

BSV infection is likely to have an adverse effect on the international exchange of *Musa* germplasm. This is particularly true of the tetraploid hybrids produced by several *Musa* breeding programs. BSV tends to show severe symptoms only in plants that are poorly managed and under stress. Generally, the more the plants are stressed, the more severe the symptoms and greater the effect on yield.

3.4. TRANSMISSION

Banana streak disease is transmitted through infected suckers upto 100% (Daniells *et al.*, 2001). The causal virus is neither sap nor soil transmitted. The virus is transmitted by mealybug *Planococcus citri* (Fig. 27) and *Planococcus* sp. in a semipersistent manner. Both these mealybugs colonize banana (Lockhart *et al.*, 1992; Lockhart, 1996).

Dysmicoccus brevipes, which are found in West Africa (Matile—Ferroro and Williams, 1995) and South America, colonize the pseudostem of banana beneath the outer leaf sheath or rhizome just below soil level, is a vector of this virus. This virus is transmitted in a semi-persistent manner. The virus is also transmitted by sugarcane mealybug *Saccharicoccus sacchari* (Kubiriba *et al.*, 2001).

Su (1998) reported that BSV was transmitted by two-mealybug bio-type (*Planococcus citri*), wild biotype with short tail (*P. citri*-ST) and a new biotype with long tail (*P. citri*-LT). Both the biotype could transmit BSV in equal fashion.

Acquisition feeding period for the citrus mealybug with BSV, seemed to be 1 day. Young instars are more efficient than old instars in transmitting the BSV. The alate adults with degenerated mouthparts cannot transmit BSV.

Daniells *et al.* (1995) reported seed transmission of BSV in cv. Mysore.

3.5. FIELD SPREAD

BSV occurs in a clonally propagated crop. Therefore the vegetative propagation is the principal method of virus spread. Plant to plant spread is through mealy bugs. Although adult mealybugs are sedentary, the early instars or "Crawlers" are highly mobile and are very efficient vectors. These mealybugs are carried by wind to neighboring plants. There is no evidence for movement of the disease into *Musa* from sources. A closely related virus, sugarcane bacilliform virus (ScBV) (Lockhart and Autrey, 1988), infects *Musa* (Bouhida *et al.*, 1993) can be transmitted from sugarcane to *Musa* by the pink mealybug *Saccharococcus sacchari*, produces symptoms in *Musa* similar to those caused by BSV (Lockhart, 1996), and occurs widely in sugarcane (Comstock and Lockhart, 1990). As yet, however, there have been no documented reports of spread of ScBv from sugarcane into adjacent plantings of banana or plantain. Mealybugs on banana in Ecuador have been seen to be managed by ants, which may move them from plant to plant, thus spreading the disease. The likely seed-borne nature of BSV and the fact that the virus can be present in germplasm without expressing symptoms have caused serious problems for some conventional breeding programme. All banana and plantain species contain segments of the DNA of BSV. The process of tissue culture can activate the integrated

DNA of the virus in the DNA of the plant, so that new plant becomes infected with virus.

3.6. VIRUS PURIFICATIONS

The virus has been purified by Lockhart (1986) and Thottapilly *et al.* (1998).

3.7. SEROLOGY

Naturally occurring isolates of BSV are characterized by high degree of both serological and genomic variability, a factor which has created considerable difficulty in developing reliable antigen-based or genome-based detection methods (Lockhart and Olszewski, 1993). To overcome this problem, polyclonal rabbit, mouse and chicken antisera were raised against a mixture of serologically distinct BSV isolates (Ndowora, 1998). Although these antisera are now used with confidence to detect BSV isolates, there always remains the possibility that there exist virus isolates, which are unrelated serologically to the pool of BSV antigens used for antiserum production. Thottapilly *et al.* (1998) have produced polyclonal antibodies against Nigerian isolate of BSV with a titre between 1:10000 and 1: 40000.

3.8. PARTICLE MORPHOLOGY

Virus particles are non-enveloped bacilliform measuring on an average 130 × 150 nm in size (Fig. 28) (Lockhart, 1986; Dabeck and Waller, 1990).

3.9. ETIOLOGY

Banana streak disease is caused by a plant Pararetrovirus having genome 7.2—8.2 kbp noncovalently closed circular ds DNA. (Lockhart, 1990; Medberry *et al.*, 1990; Lockhart and Olszewski, 1994; Lockhart *et al.*, 1998) and belongs to the genus Badnavirus. Badnaviruses are only the second group of ds DNA plant viruses to be described (Lockhart, 1990) and the virus belongs to the family Caulimoviridae. This family includes plant and animal viruses in which the virus replicates their genome DNA via an RNA intermediate using reverse transcription (Medberry *et al.*, 1990; Mayo, 2002).

3.10. HOST RANGE

Like many badnaviruses, BSV has a very restricted host range. BSV has been found naturally only in *Musa* (banana, plantain, bluggoe), and has been transmitted experimentally only to *Musa* and *Ensete*. Experimental transmission of BSV to *Musa textilis* and the related genera *Heliconia* and *Strelitzia* has been unsuccessful. The significance of the restricted host range of BSV is that spread of BSV must of necessity occur from banana to banana. There is some information on cultivars and subgroups that appear susceptible. The first reports of banana streak were from cultivars in the Cavendish subgroup (AAA) in Cote d' Ivoire (Lassoudiere, 1974) and Morocco (Lockhart, 1986). Sebasigari and Stover (1988) found banana streak disease in Kabungo. Prefecture in Rwanda in 'Ney Poovan' (AB), 'Pisang Awok' (AAB) and the East African high-land beer-making cultivars of the Lujugira-Mutika subgroup (AAA). In the Bukoba district of Tanzania, the disease was seen on 'Gros Michel' (AAA) and cooking cultivars in the AAA Lujugira - Mutika subgroup. They also reported banana streak as present in Zanzibar on 'Giant Cavendish' (AAA, Cavendish subgroup) and 'Pukusa' (AAB), syn. 'Silk'). Later Dabek and Waller (1990) recognized banana streak symptoms in 'Mjenga' (A, syn 'Sucrier'), 'Kipaka' (AA, syn. 'Paka'), 'Bukoba' (AAA, syn. 'Gros Michel'), 'Paji' (AAA), 'Mzunga Mwekundu' (AAA, syn 'Red'), Mzunga Mweupe (AAA, syn. 'Green Red'), 'Mkono Wa Tembo' (AAB, 'Horn plantain type'), 'Kikonde Kenya' (AAB, syn. 'Mysore') and 'Kijakaze' (AAB, Pome Subgroup) in a germplasm collection in Zanzibar.

Results of the tests undertaken at INIBAP Virus Indexing Centres, confirm the susceptibility of the Cavendish, plantain and Lujugira-Mutika subgroups and the widespread distribution of diseased 'Mysore'. Additional clones found with BSV particles include 'Kisubi' (AB, 'Ney Poovan') from Burundi, 'Pisang Berangan' (AAA, syn. 'Lakalan) in Malaysia, 'Pisang Raja' in India and Malaysia, cultivars in the AAB Maia maoli—popoulu subgroup from

Columbia and Australia and 'Monthan' (ABB, Bluggoe subgroup) from Cote d' Ivoire. The virus has also been found in some AAAB hybrids from breeding program in Guadeloupe, Honduras, Nigeria and Brazil.

3.11. BIO-CHEMICAL CHANGES

Peroxidase, polyphenol oxidase and phenyalanine ammonialyase activity increase in BSV infected plants (Gomathi *et al.*, 1993).

3.12. DETECTION OF BSV

Detection of BSV is problematic due to serological and genomic heterogeneity of virus isolates (Lockhart and Olszewski, 1993), the erratic appearance of symptoms (Dahal *et al.*, 1998) and the uneven distribution of the virus in plants. Diagnosis based on foliar symptoms may be very unreliable because of the sporadic nature of symptom expression throughout the year (Jones and Lockhart, 1993). Symptoms may be totally absent, or may be indistinct, under certain conditions. Symptom expression is also usually totally lacking in plants derived from *in vitro* multiplication. Attempts to develop sensitive, reliable methods for BSV indexing using antigen-based (serology) and genome-based (amplification by polymerase chain reaction, PCR) have been only partly successful. Both these approaches have been compromised by the very high degree of both serological and genomic heterogeneity that exists among isolates of the virus (Lockhart and Olszewski, 1993). The high degree of genomic heterogeneity that exists among BSV isolates can be explained by the mechanism of replication of Badnaviruses and other Pararetroviruses. The virions of these viruses contain a ds DNA which is replicated by reverse transcription via RNA intermediate. This process is inherently prone to the generation of genomic variants. Because these variants are maintained and disseminated by vegetative propagation, there is little selection pressure applied against these variants.

Lockhart (1995) mentioned that of the virions in a single banana plant are isolated and a PCR amplitude carried out, around 20 products can be obtained. If these are sequenced, they will all be different. BSV includes thousands, perhaps even millions, of different genotypes. If banana plant is to be certified as free of BSV then we must be sure that all isolates are detectable, and none of them are present.

Wetzel *et al.* (1992) and Harper *et al.* (1999a) developed an IC-PCR based strategy to detect episomal BSV in banana and plantain. Antisera used in IC-PCR are capable of binding a large number of BSV serotypes. This technique is suitable for large-scale screening of *Musa* for episomal BSV which is necessary for germplasm movement.

The presence of the virus can be detected serologically using ELISA and immunoelectron microscopy techniques (Lockhart, 1986; Lockhart and Olszewski, 1993; Ndowora and Lockhart, 1998; Singh *et al.*, 1996; Hughes, 1998; Selvarajan *et al.*, 1997; Reichel *et al.*, 1997; Thottapilly *et al.*, 1998). Tripple antibody sandwich (TAS) ELISA has been found more sensitive.

In spite of marked improvements in the reliability of BSV detection by ELISA, there remain virus isolates that are detected weakly or not at all by this method. However, these and all other BSV isolates can be detected by immunosorbent electron microscopy (ISEM) using partially purified leaf-tissue extracts (Bouhida *et al.*, 1993).

Immunosorbent electron microscopy of viral minipreps, using a multi-strain antiserum to BSV (Diekmann and Putter, 1996; Thottapilly *et al.*, 1998), appears to be the most reliable method for detection of a variety of strains of BSV. This combined assay is more sensitive than either assay alone. The commercial AGDIA ELISA kits for BSV and ScBV are only able to detect some isolates of BSV.

Electron microscopy of purified virus preparation is another reliable method for the detection of BSV (Singh *et al.*, 1996; Hughes, 1998). Thottapilly *et al.* (1997), developed immunocapture-polymerase chain reaction (IC-PCR) amplification for the detection of BSV. This method avoids false positives due to integrated viral sequences and is a potentially useful tool in diagnosing BSV infection.

3.13. CLONING, SEQUENCING AND PRESENCE OF BSV GENOMIC SEQUENCES IN *MUSA* DNA

Harper and Hull (1998) cloned and sequenced the genome of a Nigerian isolate of BSV and shown to comprise 7389 bp and to be organized in a manner characteristic of Badnavirus. PCR with primers based on sequence data indicated that BSV sequences are present in the banana genome.

La Fleur *et al.* (1996) and Lockhart (1996) reported that BSV genomic sequences are integrated into the *Musa* genome, and are capable of giving rise to episomal BSV infection and disease expression under certain conditions, one of which tissue culture stress, has been identified. This previously undescribed phenomenon has some important practical implications for banana improvement by conventional breeding. This discovery led naturally to the speculation that *de-novo* virus infection might arise from integrated viral sequences even though no parallel phenomenon has previously been documented for either plant or animal Pararetroviruses. If the hypothesis proved to be correct and if it were possible to identify potentially functional integrated viral sequences, then it would be possible, over the short term to identify *Musa* genotypes which plant breeder could use without danger of BSV infection arising in the progeny. Over the longer term, it would be necessary to understand the cellular and molecular mechanisms through which integrated badnaviral sequences are expressed episomally, with a view to developing strategies which might interfere with this process.

To confirm that the badnavirus-specific sequences being amplified were indeed contained with the *Musa* genome and not simply associated with it, hybridization analyses were carried out, using the labeled PCR product as a probe. This showed that in virus-free plants, the label was associated with DNA of a molecular weight (>20kb) much greater than that of virion DNA. Similar hybridization following restriction enzyme digestion of cellular DNA showed that label was associated only with some fragments, confirming the specificity of the hybridization signal (Lockhart *et al.*, 1998). It was therefore hypothesized that :

1. All *Musa* genotypes contain integrated badnavirus sequences (Integrated form of BSV).
2. Under Certain conditions (e.g. tissue culture), virions (episomal form of BSV) are produced from the integrated form and disease symptoms develop.

If the above hypothesis is correct, then the following questions arise :

1. What is the nature of the integrated viral sequences?
2. What is the relationship between the integrated and episomal forms of BSV occurring in different *Musa* genotypes?
3. Does any correlation exist between genotype and/or pedigree and presence of episomally-expressible BSV integrants?

To characterize integrated BSV sequences in *Musa*, Lockhart *et al.* (1998) constructed genomic libraries of Calcutta 4 (AP) and Obino L' Ewai (AAB) and used two parental genotypes in breeding program to obtain improved tetraploid plantain (AAAB). Calcutta 4 genomic clones were screened by hybridization with a probe derived by PCR amplification using badnavirus specific degenerate oligonucleotide primers (La Fleur *et al.*, 1996). When the Obino L' Ewai genomic library was screened using a probe consisting of a Pst I fragment of genomic DNA from episomal virus isolated from a tetraploid Obino L' Ewai × Calcutta 4 hybrid, integrated BSV sequences were identified in 11 genomic clones (Ndowora, 1998). Sequence analysis revealed a high degree (>90%) of homology between viral sequences integrated in Obino L' Ewai and that of the episomal virus occurring in the tetraploid hybrid progeny derived from it. Interestingly, genomic DNA of Calcutta 4 did not hybridize with the probe derived from genomic DNA of the episomal virus occurring in the tetraploid hybrid of which it was the male parent.

The episomal virus might be derived from one or more integrated viral sequences, which were largely similar across a range of genetically diverse *Musa* genotype. Additional evidence to support the hypothesis that episomal virus appearing in tissue culture derived plants originated from integrated viral sequences was obtained from nucleotide sequence comparisons between the integrated and episomal forms. It was found that a BSV sequence integrated in Obino L' Ewai shared > 99% sequence homology with genomic DNA of episomal virus isolated from one of its tetraploid progeny which had been propagated *in vitro*. Once again this is in contrast to naturally occurring BSV isolates, which are characterized by a high degree of genomic heterogeneity (Lockhart and Olszewski, 1993).

Sequencing of clones of plantain cv. Obino L' Ewai genomic DNA (Harper *et al.*, 1999) revealed an interface between banana streak badnavirus and *Musa* sequence and a complex BSV. *In situ* hybridization revealed 2 different BSV sequence locations in chromosomes and a complex arrangement of BSV and *Musa* sequences was revealed by probing stretched DNA fibres. This is thought to be the first report of integrated sequences that possibly lead to a plant pararetrovirus episomal infection by a mechanism differing markedly from animal retroviral system.

Ndowora *et al.* (1999) reported that BSV DNA was integrated into the genome of banana and plantain *Musa* spp. although BSV does not employ integration during its infection cycle. From the arrangement of the integrated BSV DNA it is suggested that it can yield an infectious episomal genome via homologous recombination. This is thought to be the first report of integrated DNA of a non-integrating virus giving rise to an episomal viral infection. Tissue culture is identified as a possible trigger for the infection.

3.14. RELATIONSHIP WITH OTHER BANANA VIRUSES

Moroccan banana streak disease appears to be similar or identical to the dotted-line mosaic (Mosaiaque an tirets) of banana reported from the Ivory Coast (Lassoudiere, 1974, 1979). Symptoms similar to BSV have been observed in the Canary Islands and in Jordan. However, presence of BSV in these countries has not been confirmed. BSV appears to belong to a group of virus that includes Cacao

Swollen shoot (Adomako *et al.*, 1983), Colocasia bacilliform virus (James *et al.*, 1973), Rice tungro bacilliform virus (Omura *et al.*, 1983) and others.

Sugarcane bacilliform virus (ScBv) has been transmitted by the mealy bug *P. citri* and *S. sacchari* from infected sugarcane to healthy Dwarf Cavendish' banana, which develope symptoms identical to those caused by BSV. Many isolates of ScBV are serologically related to isolates of ScBV (Lockhart and Autrey, 1988, Lockhart and Olszewski, 1993). BSV does not infect sugarcane (Lockhart, 1995).

3.15. HOST-VIRUS INTERACTIONS

It has been noted in the past that tetraploid hybrids seem to be more "susceptible" to BSV. Studies at International Institute of Tropical Agriculture (IITA) Nigeria on micro-propagated banana and plantain hybrids confirmed that plantain hybrids had a high incidence of symptomatic plants which were negligible or absent in their parental lines and other landraces (Dahal *et al.*, 1997a). In addition, it has been noted that while plantain hybrids express symptoms under cooler (22°C) as well as ambient temperatures, most cooking bananas did not express symptoms under either regime although virus could be detected by ISEM in some of the plants (Dahal *et al.*, 1997 b; Dahal *et al.*, 1998).

3.16. MANAGEMENT

There are currently no really effective ways of controlling episomal BSV apart from ensuring that planting material is disease free and by eradication. There is conflicting evidence on the extent and significance of horizontal vector-borne spread of BSV by mealy bugs. Thus, there is a need to gain firm data on this and on whether BSV can spread from sugarcane to banana. As there appear to be few or no natural forms of resistance to BSV which could be bred in by conventional crossing, possible transgenic approaches should be considered. Although such approaches are being used widely and successful with other virus-host combinations, the potential complications of integrated viral sequences, whether active or inactive, have to be considered in the case of BSV. Similarly, there is no case history which can show the potential of the transgenic approach in overcoming the active integrated form. Suckers are more resistant to BSV infection as compared to tissue culture raised plantlets. Greenhouse studies have shown that tolerance to BSV is correlated with the size of banana plantlets. For example, when banana plantlets 5-10 cm tall were inoculated with BSV, the infection rate reached 75% while the average incubation period was 22 days. When plantlets 30 cm tall were inoculated with BSV, the corresponding figures were 50% infected, with an incubation period of 80 days. For plantlets 50 cm tall, only 25% were infected after an incubation period of 135 days.

Phytosanitation

Roguing is traditionally used to remove plants which are apparently unhealthy so that healthy plants can be substituted. In case of BSV infected plants, severely affected plants may not produce acceptable yield. It can therefore be useful to remove plants exhibiting severe symptoms and replant with healthy 'Suckers'. It must be noted however, that symptomless suckers can be obtained from mother plants showing severe symptoms, and that suckers with symptoms can grow from apparently symptomless mats. In the former case, the symptomless suckers are virus infected and may later develop symptoms.

Good Management

Good farm management appears to allow plants to attain a high percentage of their yield potential, despite infection with BSV. The major factors which seem to reduce symptom expression and thus have subsequent effect on yield are: reduced water stress, mulching, reduced weed competition and minimal stress from other pests and diseases. Mulching is likely to be effective in reducing the effect of BSV on the plants by helping to reduce water loss and therefore also reducing water stress.

4

BRACT MOSAIC OF BANANA

4.1. GEOGRAPHICAL DISTRIBUTION

Philippines, Uganda, Ghana, Nigeria, Zanzibar, Malawi, India and Sri Lanka (Magnaye and Espino, 1990; Pasberg-Gauhl *et al.*, 1995; Vuylsteke *et al.*, 1996; Singh *et al.*, 1996; Rivera *et al.*, 1993; Diekmann and Putter, 1996; Thomas and Magnaye, 1996).

Bract mosaic is a newly recognized disease of banana in the Philippines, Island of Mindanao although evidence suggests that it has been present for many years. This disease was first recorded by Magnaye and Espino (1990) and was first observed on cultivars, Canara, Pisung, Maduranga and Morado in 1979 in Davao City, Mindano. Bract mosaic has also been observed in Ghana and Nigeria (Pasberg-Gauhl *et al.*, 1995), Malawai (Vuylsteke *et al.*, 1995) and India (Singh *et al.*, 1996). A detailed account of banana bract mosaic virus (BBrMV) has been given by Ploetz (1994).

The disease was given the name bract mosaic in a meeting of banana virologists that took place in Los Banos in 1988 and included in a list of viruses of quarantine importance (Frison and Putter, 1989).

4.2. SYMPTOMATOLOGY

Symptoms are most noticeable in the bracts where discoloured streaks are evident; mosaic in the pigmentation of pseudostems after the removal of dead leaf sheaths are also distinctive (Fig. 29). Suckering is also suppressed and suckers that do emerge are distorted and deeply pigmented (Fig. 30). Chlorotic streaks parallel to veins can sometimes be seen in leaves and petioles (Fig. 31).

Fig. 29

Fig. 30

Fig. 31

Fig. 32

Fig. 33

Fig. 34

Fig. 35

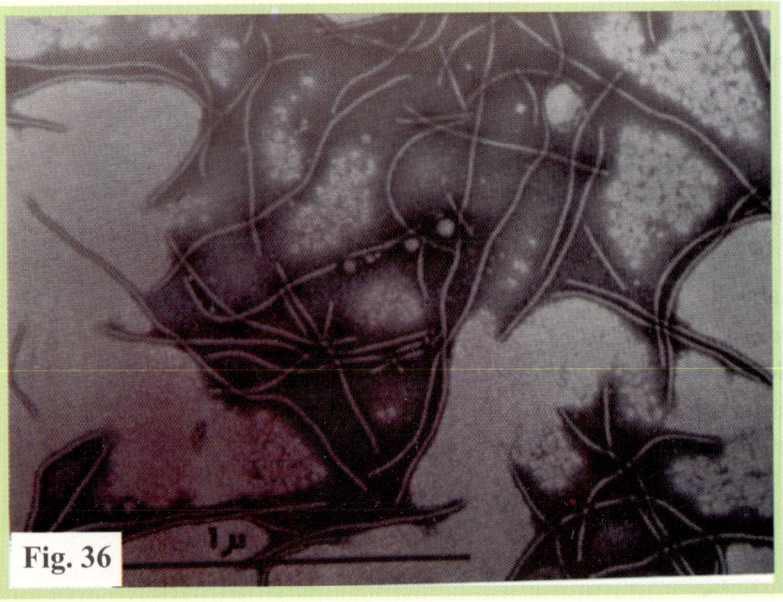

Fig. 36

Symptoms most often seen in Cavendish are a pronounced purple mosaic or streaking on the pseudostem (Fig. 32, 33). Plants become spindly and die prematurely (Fig. 34, 35).

4.3. ECONOMIC IMPORTANCE

Banana Bract mosaic virus (BBrMV) has been reported to cause considerable damage in Cavendish plantations. In India banana bract mosaic is mostly present in states of Tamil Nadu and Kerala. However, this has also been noticed in the states of Karnataka, Maharashtra, Gujarat, Bihar and Assam. In Tamil Nadu, the percentage incidence of BBrMV in Nendran (AAB), Poovan (AAB), Robusta (AAA), Red banana (AAA), Ney Poovan (AB), Monthan (ABB), Pisang Awak (ABB) were 15.93, 28.23, 4.57, 56.80, 5.31, 36.12, 0.5 respectively (Selvarajan and Singh, 1997). Variety Nendran is more susceptible to BBrMV. The disease is lethal in Cavendish cultivars and of great concern to quarantine. This disease caused a part yield loss upto 40% in highly susceptible varieties, Saba and Cardaba (Espino *et al.*, 1990). Yield losses of upto 40% were noted in "Cardaba" and "Lakatan" (AAA) varieties (Kenyon *et al.*, 1996; Thomas and Magnaye, 1996). The failure of fruit to fill on infected plants has been noted in India (Diekmann and Putter, 1996). Cherian *et al.* (2002) reported 70% yield reduction in variety Robusta followed by 52.5% in variety Nendran.

4.4. TRANSMISSION

BBrMV is transmitted through infected suckers, corms and micro propagated plantlets.

This virus has also been experimentally transmitted to healthy bananas by the at least three aphids; *Rhopalosiphum maidis* and *Aphis gossypii* (Magnaye and Espino, 1990) and *Pentalonia nigronervosa* (Munez, 1992; Diekmann and Putter, 1996). *P. nigronervosa* transmitted BBrMV after an acquisition access period of 1 min, indicating that transmission is of the non-persistent type (Munez, 1992). Efficiency of transmission with the latter species was less than 10% (Caruana and Galzi, 1998).

Attempts to transmit BBrMV by sap inoculation to herbaceous indicator plants have so far been unsuccessful (Magnaye and Espino, 1990; Munez, 1992). However, occasional sap transmission from banana to banana has been achieved (L.V. Magnaye and L. Herradura, Davao, 1998, unpublished).

4.5. VIRUS PURIFICATION

BBrMV has been purified by Espino *et al.* (1990) using the procedure followed by Retuerma (1981). The purified preparations showed an absorbance ratio of 1.25 at 260/280 nm which indicated higher nucleic acid than that of protein. This virus has also been purified by Rodoni *et al.* (1997) and Thomas *et al.* (1997).

4.6. ETIOLOGY

Banana bract mosaic is caused by a potyvirus (Potyviridae) (Munez, 1992; Bateson and Dale, 1995; Singh *et al.*, 1996; Rodoni *et al.*, 1997; Thomas *et al.*, 1997).

4.7. PARTICLE MORPHOLOGY

Flexuous rod shaped particles has been detected in diseased tissues (Espino *et. al.*, 1990). Singh *et al.* (1996) observed large number of flexuous rod shaped virus particles (720 × 15 nm in size) (Fig. 36) in dip preparations from symptomatic bracts of banana.

4.8. SEROLOGY

Both, polyclonal and monoclonal antisera against BBrMV have been prepared (Espino *et al.*, 1990) and now they are commercially available.

4.9. SEROLOGICAL RELATIONSHIP

Isolates of BBrMV from India, Sri Lanka and the Philippines are all closely related serologically (Thomas *et al.*, 1997). Weak serological relationship have been demonstrated between BBrMV and other potyviruses, including abaca mosaic, dasheen mosaic, maize dwarf mosaic, wheat streak mosaic, sorghum mosaic and sugarcane mosaic viruses (Thomas *et al.*, 1997).

4.10. DETECTION

Both serological and nucleic acid-based assay are now available for BBrMV. The presence of BBrMV can be detected by serology using ELISA technique (Espino *et al.*, 1990; Singh *et al.*, 1996; Thomas *et al.*, 1997) and also by electron microscopy of dip preparations (Singh *et al.*, 1996). The virus can also be detected by nucleic acid probe specific of BBrMV (Reddy *et al.*, 1996; Thomas *et al.*, 1997). The virus can also be detected by PCR in total nucleic acid extracts from infected plants, using either virus specific or degenerate potyvirus group primers (Bateson and Dale, 1995; Thomas and Magnaye, 1996; Thomas *et al.*, 1997).

4.11. VARIETAL REACTION

The disease is quite common in varieties, Saba, Cardaba and Abuhon Cooking banana in Philippines. Other cultivars namely Turankong Viante, Cohol, Dukdakan, Pulutan, Robusta, Ambon, Tukoroo including wild *Musa balbisiana* is susceptible. Serological tests using polyclonal antibodies with DAS ELISA, revealed that Abuhon and Cardaba are highly susceptible which 'K' Thong Det 52' was the least infected of all banana varieties (Espino *et al.*, 1990). The virus has been detected in a wide range of naturally infected banana cultivars and genotypes, and no resistance to the virus has been noted.

In India varieties Nendran, Robusta, Ney poovan and Poovan are susceptible to BBrMV (Thangavelu and Singh, 1996). However, symptoms have also been observed in other cultivars too. BBrMV has been reported to infect abaca (*Musa textilis*) under natural conditions (Sharman *et al.*, 2000).

4.12. PROPERTIES OF THE VIRUS

Purified virions contain a major coat protein of 38K-39K Da protein as estimated by sodium dodecylsulphate polyacrylamide gel electrophoresis (SDS-PAGE) which reacted with a general potyvirus antiserum in Western blots (Bateson and Dale, 1995). Virions have a buoyant density in calcium chloride of 1.29—1.31 g/ml and an A 260/280 = 1.17. Three major protein species with size of 31, 37 and 39 Kda were resolved from dissociated virions, and all reacted positively with polyclonal antibodies of BBrMV. Infected cells contained typical pinwheel inclusions (Thomas *et al.*, 1997).

Bateson and Dale (1995) degenerated PCR primers to firstly determine whether the BBrMV agent is serologically related to the potyviruses and secondly to obtain amplified genome sequences for comparisons.

Nucleotide sequence analysis indicates that BBrMV is a unique, novel potyvirus. The sequence of the 3' end of the Nlb gene, the coat protein gene and the 3' untranslated region is available for several isolates (Bateson and Dale, 1993, 1996; Rodoni *et al.*, 1979; Thomas *et al.*, 1997).

The C-terminal half of BBrMV coat protein gene amino acid sequence is most similar to that of the potyvirus MDMV (71 %) and to a lesser extent to two other potyviruses, SCMV (70.7%) and OMV (68%). In the 3′ un-translated region, BBrMV is most similar to OMV (40%). The PCR products of partially purified BBrMV contained a 5' open reading frame of upto 150 amino acids and a 3' untranslated region of upto 190 nucleotides which were clearly related to the C-terminal half of the coat proteins and the 3' unstructural regions, respectively of potyvirus. These results indicated that BBrMV is a distinct potyvirus and is distinct from any other potyvirus for which sequence information is available. Further, it would appear that BBrMV is not a strain of SCMV. Rodoni *et al.* (1997) characterized BBrMV present in India by sequence homology and nucleic acid assay hybridization. There was 98.7% sequence homology between *M. balbisiana* and Klue Terapod isolates from Coimbatore. The Pisang Awak isolate from Tiruchirapalli (T.N.) has 95.9% homology with the *M. balbisiana* isolate. When compared with the Philippines isolate of BBrMV the *M. balbisiana*, Klue Terapod, and Pisang Awak isolates showed 97.6, 96.6 and 99% homology, respectively. Amino acid homology between Klue Terapod and *M. balbisiana* (97.9%) were lower than the nucleotide sequence homology (98.7%) and were due to base changes at positions 158, 234, and 339 that resulted in 3 amino acid changes at positions 53, 78 and 113. Sequence comparisons of the C. terminus of the coat-protein-coding and 3' untranslated regions revealed that the Indian isolates of BBrMV had greater than 96.6 and 97.2% homology with a Philippines isolate at the nucleotide and amino acid levels, respectively.

4.13. MANAGEMENT

The disease spread can be minimized by immediate treatment of infected and survived plants with aphidicides, complete destruction of infected plants and weed control. The use of indexed, virus-free planting material is the best means of control.

5

BANANA MOSAIC—TOBAMOVIRUS

5.1. SYMPTOMATOLOGY

This disease was first reported from Bangalore (India) by Singh (1988b). The characteristic symptoms are mosaic or discontinuous linear or spindle-shaped streakings on the leaf, either generally distributed over the leaf lamina or in bands of half an inch or more in width, extending from margin to midrib (Fig. 37). Severely infected leaves exhibit greenish-yellow mottled look. Symptoms are also apparent on the narrow leaves of young suckers or exised buds. On young leaves, the streaks or linear splashes look whitish, pale yellow or yellowish-green. The infected plants remain stunted and do not bear fruits (Singh, 1988b).

5.2. TRANSMISSION

The virus is transmitted by mechanical sap inoculation from banana to banana and other herbaceous hosts. Symptoms on banana appear 45-60 days after sap inoculation. The virus is not transmitted by aphids, or whiteflies.

5.3. HOST RANGE

On artificial inoculations by sap, the virus can infect following plant species: *Lycopersicon esculentum, Nicotiana rustica, N. tabacum* cv. W.B., *Cucumis melo, Datura stramonium, Capsicum annuum,* and *Gynura aurantifolia.* These hosts produce mosaic-mottling symptoms

However, *Chenopodium amaranticolor, C. quinoa* and *N. glutinosa* produce necrotic local lesions on inoculated leaves.

5.4. PHYSICAL PROPERTIES

The thermal inactivation point of the virus lies between 85-90°C, dilution end point between 10^{-4} —10^{-5} and ageing *in vitro* upto 24 hr at room temperature (25-32°C).

5.5. SEROLOGY

The virus is highly immunogenic and is serologically related to tobacco mosaic virus (TMV).

5.6. PARTICLE MORPHOLOGY

Electron microscopy of leaf dip preparations revealed the presence of rigid-rod shaped virus particles measuring on an average 300 × 15 nm in size (Fig. 38), in high concentrations (Singh, 1988b).

5.7. ETIOLOGY

Based on host reaction, physical properties, serology and electron microscopy, the banana mosaic is caused by a strain of TMV (Tobamovirus)(Singh,1988b).

5.8. MANAGEMENT

The spread of the virus can be checked by selection of virus-free planting material. Infected plants must be rouged out from the field to avoid further spread.

6

BANANA MILD MOSAIC VIRUS

Uncharacterized, filamentous, virus-like particles have been noted frequently in *Musa* germplasm accessions and field specimens (Rivera *et al.*, 1992; Anon 1993; Lockhart, 1994; Belacazar *et al.*, 1998; Caruana and Galzi, 1998). Research results suggest that in most cases the same single virus is present.

The economic importance of this virus is not known, nevertheless, the yield may be affected in both banana and plantain under some conditions.

This disease appears to occur widely, wherever bananas are cultivated, and extensively in *in vitro* germplasm collections. Field records exist from Australia, Africa, South and South East Asia, Central and South American and the Caribbean, in a wide variety of *Musa* cultivars and genotypes.

6.1. SYMPTOMATOLOGY

The symptomatology of the disease, which has been named banana mild mosaic virus (BanMMV), is uncertain. In many cases, both in glasshouse-grown plants, and in the field, symptomless infection occurs. 'Ducasse' (ABB, syn. 'Pisang Awak') in the field has been observed with chlorotic mosaic and streak symptoms (Fig. 39). These symptoms which are also often seen on young 'Ducasse' plants derived from infected tissue cultures, may disappear as the plant matures. Transitory chlorotic leaf streaks has been observed on some leaves of 'Daluyao' (AAB, plantain subgroup) infected with BanMMV alone and growing under glasshouse conditions. 'Pisang Seribu' (AAB) infected with BanMMV alone, displays silvery streaks on the leaf lamina, especially on the distal part, which are best

Fig. 37

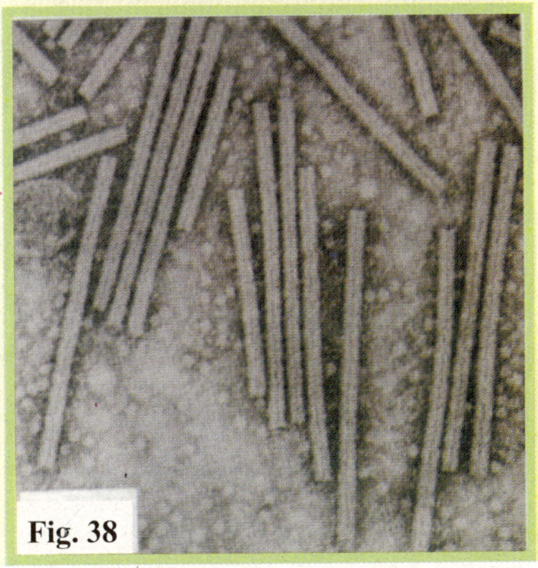

Fig. 38

Fig. 39

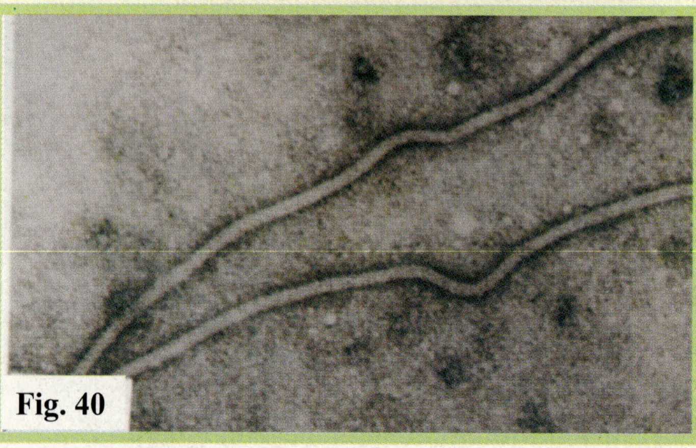
Fig. 40

viewed from the underside of the leaf. Plants of 'Gros Michel' (AAA) from Latin America have displayed large chlorotic blotches on the leaves, stunting and delayed bunching. Mixed infections of BanMMV with BSV and CMV are common. In case of BanMMV and BSV infection, symptoms are reminiscent of those caused by BSV alone. In mixed infection with CMV, the presence of BanMMV seems to correlate with an additional leaf necrosis symptom (Caruana and Galzi, 1998).

6.2. ETIOLOGY

Virus particles are flexuous filaments about 580 nm long and 14 nm wide (Fig. 40). The coat protein has an apparent Mr of 30 KDa in SDS-PAGE and, translated from the coat-protein gene sequence, is estimated to be 27 KDa. The concentration of virions in infected plants is variable, though generally low with yields of purified virions in the order of 1 mg kg^{-1} tissue.

The genome is ssRNA of size c. 7.4 kb, contains five ORFs (from the 5' end: ORF1/polymerase, probably 205 KDa, triple gene block, 26, 12 and 8 KDa and ORF 5/coat protein 27 KDa), 3'and 5' untranslated regions and a poly-A tail. A ds RNA of c.7 kbp has been detected in plants of 'Poingo' (AAB, Maia maoli-popular subgroup) and 'Kelong 'McKintu' (AAB, plantain subgroup) containing BanMMV particles. BanMMV shares some, but not all, properties with the potex virus group. It is serologically related to members of that group. It has also a genome organization and particle length consistent with that group.

Gambley and Thomas (2001) have sequenced the complete genome of BanMMV and found that the genome consists of 7352 nucleotide, encoding five ORFs of 20.5, 25.5, 12.4, 8.0 and 26.8 KDA. This virus is closely related to fovea and Carlaviruses.

6.3. DETECTION

The presence of the virus can be detected by serology and electron microscopy. ELISA and ISEM techniques have been found useful in virus detection. Monoclonal antibodies have been prepared and these have been used in tissue blot and western blot assays (Caruana *et al.*, 1995). A triple antibody sandwich ELISA has been developed which uses polyclonal coating antibodies and these MAbs as detecting antibodies. PCR primers have been developed which allow detection of BanMMV in both total nucleic acid extracts from infected plants and in immunocapture reverse transcriptase (RT) PCR, using crude sap extracts.

6.4. DISEASE CYCLE AND EPIDEMIOLOGY

The disease spread through vegetative propagation (both conventional planting material and tissue culture). This virus is one of the most common contaminants of international germplasm collections, probably due to both its wide distribution and the frequently symptomless nature of infection.

6.5. TRANSMISSION

BanMMV has not been transmitted mechanically with sap extracts or partially purified preparations to banana or herbaceous indicators. This virus is not transmitted by banana aphid, *Pentalonia nigronervosa*. Under natural conditions in Guadeloupe, plants established from virus-free tissue cultures of cultivars in the Cavendish subgroup (AAA) subsequently became field-infected with CMV and BanMMV, frequently as mixed infections, and the incidence of infection increased during the cropping cycle. This implies that natural transmission of BanMMV was occurring generally simultaneously with the aphid transmitted CMV (Caruana and Galzi, 1998).

6.6. HOST REACTION

All *Musa* genotypes appear to be susceptible to BanMMV, though this observation is based only on natural infections. All attempts to transmit the virus to banana or to herbaceous indicators have been unsuccessful (Caruana and Galzi, 1998).

6.7. MANAGEMENT

The use of indexed, virus free vegetative planting material is recommended.

7

BANANA DIEBACK VIRUS

Hughes *et al.* (1998) reported a new virus disease namely "banana die-back" from Nigeria. The important symptoms are leaf crinkling, leaf necrosis and cigar leaf dieback in cv. Valery. Subsequent suckers from the same mats were progressively more stunted. A 28 to 30 nm isometric virus was purified from the infected plants and antiserum was produced. Mechanical inoculation with partially purified virus preparations resulted in stunting and development of pinpoint chlorotic lesions on *Vigna unguiculata* TVu - 76 and symptomless systemic infections on *Nicotiana occidentalis*. The virus is not mechanically transmitted from *N. occidentalis* to banana. A serological relationship between this virus, banana die-back (BDBV) and tobacco ring spot, tomato ring spot and Cacao necrosis nepovirus was found. The virus is not transmitted by nematodes.

8

SEED-BORNE MUSA VIRUSES

Seedlings of *M. balbisiana* and *M. acuminata* from Central America developed virus symptoms when grown in the glasshouse at Berkeley, California (Gold, 1972). The virus is cucumber mosaic and is probably seed transmitted through pollen. About 10% of the seedlings of *M. balbisiana* grown in Honduras showed virus symptoms consisting of a white streaking of the leaves. Severely affected seedling died but many subsequently produced symptomless leaves and appeared to recover. Seedlings of *M. acuminata* and hybrids also showed virus symptoms.

9

OTHER VIRUSES ON BANANA

Several other viruses that are either latent or have undetermined symptom expression have been recorded in banana. A latent 700 nm rod-shaped virus has been reported from Taiwan. Two filamentous viruses have been reported in Costa Rica, a 512 nm rigid rod and a 900 nm flexuous rod though the symptoms, they cause, if any, are uncertain.

REFERENCES

1. **Adam, A.V. (1962).** An effective program for the control of banana mosaic. Phytopathology, **52**: 286 (Abstr.).
2. **Adomako, D., Lesemann, D.E. and Paul, H.L. (1983).** Improved methods for the purification and detection of cacao swollen shoot virus. Ann. Appl. Biol., **103**: 109-116.
3. **Alagiamanavalan, R.S., Selvaraj, P., Veernath, L., Balakrishnan, R. and Raman, K.R. (1973).** The incidence of 'Bunchy top' infection in the flowering stage of banana. South Indian Hort., **21**: 89-93.
4. **Allam, E.K., El-Afifi, S.A., El-Ahdal, A.E. and Sadik, A.S. (1995).** Transmission and control of banana mosaic virus in Egypt. Annals of Agric. Sci. (Cairo), **40**: 525-538.
5. **Allen, R.N.(1978a).** Epidemiological factors influencing the success of roguing for the control of bunchy top disease of banana in New South Wales. Aust. J. Agric. Res., **29**: 535-544.
6. **Allen, R.N. (1978b).** Spread of bunchy top disease in established banana plantations. Aust. J. Agric Res., **29**: 1223-1233.
7. **Allen, R.N. (1983).** Spread of banana bunchy top and other plant virus diseases in the time and space. In: R.T. Plumb and T.T. Thresh (eds.). Plant Virus Epidemiology, pp. 51-59.
8. **Allen, R.N. (1994).** Banana bunchy top on screen: an introduction to disease control using sanitation measures practiced in New South Wales, Australia but not adopted for use in tropical and sub-tropical countries Inderoopilly, Australia Queensland Dept. of Prim. Industry, Land Resources Branch (1994). IV + 17 ISBNO - 7242-5437-9.

9. **Allen, R.N. and Barnier, N.C. (1977).** The spread of bunchy top disease between plantations in the Tweed River district during 1975-76. N.S.W. Dept. Agric. Biology Branch, Plant Dis. Survey (1975-76): 27-28.

10. **Anonymous (1962).** Relatrio de 1962, Vol. 1 lnst. de Investigcao agronomica de Angola, Nova Lisboa.

11. **Anonymous (1979).** *Pests and Diseases of Agricultural Crops in China*, Vol. 2. Agricultural Publishers. Peking.

12. **Anonymous (1983).** Annual Report. All India Co-ordinated Fruit Improvement Project, Cell I. Citrus, Banana, Pineapple and Papaya.

13. **Anonymous (1984).** Annual Report 1982-83. Mahatma Phule Agric. Univ., Dhule Nagar, Rahuri, Ahmednagar.

14. **Anonymous (1992).** *Bananas, Plantain* and INIBAP, *Annual Report*, 1992, INIBAP, Montpellier, France 31 pp.

15. **Anonymous (1993).** Risks involved in the transfer of banana and plantain germplasm. *In: Bananas, Plantain* and INIBAP, *Annual report* 1993. INIBAP, Montpellier, France, pp 39-97.

16. **Anonymous (1999).** Statistics from FAO website http://apps. fao. org./default. htm.

17. **Avgelis A. (1987).** Cucumber mosaic on banana in Crete. J. Phytopath.,**120**: 20-24.

18. **Barbosa, C. De J., Cumani, L.A.L., Roes, B., Oliviera, R.P. and Silva K.M. Da (2000).** Cucumber mosaic virus identification on banana plants in the North Parana State. Revista Brasiliera de Fruticultura, **20**: 112-114.

19. **Bateson, M.F. and Dale, J. L. (1995).** Banana Bract Mosaic Virus: Characterization using potyvirus specific degenerate PCR primers. Arch. Virol.,**140**: 515-527.

20. **Beaver, R.G. (1982).** Use of Picloram for eradication of banana diseased with bunchy top. Plant Dis., **66**: 906-907.

21. **Becquer, A. and Bencoino, I. (1974).** Vectors of banana mosaic virus. Revista de Agriculture, **7**: 24-34.

22. **Belalcazer, S., Reichel, H., Perez, R., Gutierrez, T., Munera, G. and Arevalo, E. (1998).** Banana streak badnavirus infection in *Musa* plantations in Colombia. In: Frison E. and Sharrock, S.L. (eds) (1998). *Banana Streak Virus: A unique Virus- Musa Interaction*? Proc. of a workshop of the PROMUSA Virology Working Group held in Montpellier, France,19-21 January, 1998. INIBAP, Montpellier, France pp. 55-57.

23. **Berg, L.A. and Bustamante, M.(1974).** Heat treatment and meristem culture for the production of virus-free bananas. Phytopathology, **64**: 320-322.

24. **Bhaktavatsalu, C.M., Sathimoorthy, S. and Madhava Rao, U.N. (1980).** Bunchy top, the phenomenon of latency, partial recovery and mild chronic infection. Proc. National Seminar on Banana Production Technol. (Abstr.) p. 27.

25. **Bird, J. and Wellman, F.L. (1962).** A mosaic disease of Musacious crops in Puerto Rico. Phytopathology. **52**: 286.

26. **Bondok, A., El-Din, T.M.N., Gabr. M.F. and Tadros, M. (1987).** Production of virus free "Hindi" banana plants utilizing meristem culture and thiouracil treatments. Assiut. Jour of Agric. Sci., **18**: 167-185.

27. **Bouhida, M. and Lockhart, B.E.L. (1990).** Increase in importance of cucumber mosaic virus infection in greenhouse—grown bananas in Morocco. Phytopathology, **80**: 981.

28. **Bouhida, M., Lockhart, B.E.L. and Olszewski, N.E (1993).** An analysis of the complete nucleotide sequence of a sugarcane bacilliform virus genome infectious to banana and rice. Jour. of Gen. Virol., **74**: 15-22.

29. **Brooks, F.E. (1999).** Incidence and distribution of bunchy top virus in American Samoa. Jour. of South Pacific Agric., **6**: 25-28.

30. **Brunt, A.A., Molyneuz, S. and Lockhart, B.E.L. (1987).** Banana streak badnavirus in viruses of Plants. Descriptions

and Lists from Video Database Cab International, U.K. 153 pp.

31. **Bryce, G. (1928).** The Bunchy top Plantain Disease. Dept. Agic. Ceylon, Leaflet No. **18**: 2.
32. **Buddenhagen, I.W. (1968).** Banana diseases in the Pacific areas. Plant Prot. Bull. F.A.O.,**16**: 17-31.
33. **Burns, T.M., Harding, R.M. and Dale, J.L. (1995).** The genome organization of banana bunchy top virus: analysis of six ssDNA components. Jour. Gen. Virol., **76**: 1471-1482.
34. **Burt, J.R. (1988).** Carnarvon's horticultural industry. J. Agric. W. Austr., 4th Ser. **29**: 27-31.
35. **Calinisan, M.R. (1938).** The three destructive disease of abaca in Davao (Bunchy top, mosaic and the vascular disease) and their control. Philippi. J. Agric., **9**: 329-333.
36. **Campbell, J.G.C. (1926).** Annual Report of the Fiji Department of Agriculture for the year 1925. pp.12-14.
37. **Capoor, S.P. (1959).** I. Bunchy top disease of banana. II. Banana mosaic, In: Proc. Seminar Dis. Hort. Plants. ICAR, New Delhi. pp. 163-172.
38. **Capoor, S.P. (1967).** Virus diseases of fruits. Indian Hort., ll: 64-69.
39. **Capoor, S.P. and Varma, P.M. (1968).** Banana mosaic in the Deccan. Indian Phytopath. Soc. Bull., **4**: 11-14.
40. **Capoor, S.P. and Varma, P.M. (1970).** Investigations on a mosaic disease of banana in the Deccan. Indian Phytopath., **23**: 62-68.
41. **Caruana, M.L. and Galzi, S. (1998).** Identification of uncharacterized filamentous viral particles of banana plants. Acta Hort., **490**: 323-335.
42. **Caruana, M.L., Galzi, S., Sechet, H., Bousalem, M. and Bringaud, C. (1995).** Etiologie de la maladie de la mosaique des practees des bananiers pour un-diagnostic cible. In:

Rapport d' Activites, 1994-1995. Laboratore de Phytovirologie des Regions Chaudes, CIRAD, ORSTOM, Montpellier, pp. 6-9.

43. **Castano, M., Galvez, G.E., Arroyave, J.A., Velasco, A.C. and Morales, F.J. (1994).** Isolation of a Columbian strain of mosaic virus of banana. Fitopatologia Columbiana, **18**: 130-134.

44. **Castellani, E. (1955).** Phytopathological problems of banana cultivation in Somaliland. Relax. Monogr. agr. Subtrop. Trop. N.S. 73-81, 13. (Rev. Appl. Mycol. **35**: 905, 1956).

45. **Castillo, B.S. and Martinez, A.L. (1961).** Occurrence of bunchy top disease of banana in the Philippines. F.A.O. Plant Protection Bull., **9**: 74-75.

46. **Cherian, K.A., Menon, R., Suma, A., Nair, S. and Sudheesh, M.V. (2002).** Effect of banana bract mosaic disease on yield of commercial banana varieties in Kerala (Abstr.) In: Global Conference on Banana and Plantain, 28-31 October, 2002, Bangalore, p. 155.

47. **Ciferri, R. (1949).** Infectious chlorosis of banana in Columbia. Nature (London), **163**: 175.

48. **Colaricco, A., Eiras, M, Vicente, M., Chagas, C.M. and Harakava, R. (1996).** Partial characterization of an isolate of cucumber mosaic cucumovirus from *Musa* sp. 'Nanicao'. Fitopatologia Bras., **21**: 268-274.

49. **Colborne, P. (1953).** Legislation for the control of bunchy top in bananas. Queensland agric. J., **77**: 85-90.

50. **Comstock, J.C. and Lockhart, B.E.L. (1990).** Widespread occurrence of sugarcane bacilliform virus in U.S. sugarcane germplasm collections. Plant Disease **74**: 530.

51. **Dabek. A.J. and Waller, J.M. (1990).** Black leaf streak and viral leaf streak: New banana diseases in East Africa. Trop. Pest Management, **36**: 157-158.

52. **Dahal, G., Hughes, J.d'A., and Lockhart, B.E.L. (1998a).** Status of banana streak disease in Africa: problems and

future research needs. Integrated Pest Management Rev., **3**: 85-97.

53. **Dahal, G., Hughes, J.d'A., Gauhl, F., Pasberg-Gauhl C., Thottapilly, A., Vuylsteke, D. and Lockhart, B.E.L.(1997d).** Evaluation of micro propagated plantain and banana hybrids for banana streak badnavirus incidence in Nigeria. Phytopathology, **87**: 522.
54. **Dahal, G., Hughes, J.d'A., Thottapilly, G. and Lockhart, B.E.L. (1997b).** Effect of temperature on symptom expression and detection of banana streak badnavirus in plantain and bananas. Phytopathology, **87**: 522.
55. **Dahal, G., Hughes, J.d'A., Thottapilly, G. and Lockhart, B.E.L. (1998b).** Effect of temperature on symptom expression and reliability of banana streak badnavirus in naturally infected plantain and banana (*Musa* spp.). Plant Disease **82**: 16-21.
56. **Dahal, G., Gauhl, F., Pasberg-Gauhl, C., Hughes, J.d'A, Thottapilly, G. and Lockhart, B.E.L. (1999).** Evaluation of micro propagated plantain and banana (*Musa* spp.) for banana streak badnavirus incidence under field and screen house conditions in Nigeria. Ann. appl. Biol., **134**: 159- 191.
57. **Dahal, G. Ortiz, R., Tenkouano, A., Hughes, J.d'A, Thottapilly, G., Vulysteke, D. and Lockhart, B.E.L. (2000).** Relationship between natural occurrence of banana streak badnavirus and symptom expression, relative concentration of viral antigen and yield characteristics of some micro propagated *Musa* spp. Plant Path., **49**: 68-79.
58. **Dale, J.L. (1987).** Banana bunchy top: An economically important tropical plant virus disease. Adv. Virus Res., **33**: 301-325.
59. **Dale, J.L.(1988).** The status of disease indexing and the international distribution of banana germplasm. Conservation and movement of vegetatively propagated germplasm *in vitro* culture and disease aspects. Rome; International Board for plant genetic Resources, pp. 34-46.

60. **Dale, J.L.(1990).** Banana and plantain Agricultural Biotechnology. Opportunities for International development. Wallingford, U.K., CAB International,1990. pp. 225-240.

61. **Dale J.L., Phillips, D.A. and Parry, J.N. (1986).** Double stranded RNA in banana plants with bunchy top disease. J. Gen. Virol., **67**: 371-375.

62. **Dale, J.L., Burns, T., Oehlschlager, S., Karan, M. and Harding, R. (1993).** Banana bunchy top virus: prospects for control through biotechnology . In: Proc. of the Workshop on Biotechnology applications for banana and plantains Improv., San Jose, Costa Rica, 27-31 Jan., 1992.

63. **Daniells, J., Thomas, J.E. and Smith, B.J. (1995).** Seed transmission of banana streak virus confirmed. Infomusa **4**: 7

64. **Darnell-Smith, G.P. (1919).** Bunchy top of banana in the Tweed River District. Agric. Gaz. N.S.W., **30**: 809-814.

65. **Darnell-Smith, G.P. (1924).** 'Some aspects of the Bunchy Top disease of bananas'. Proc. Pacific Sci. Congr., 1923, 149-153.

66. **Darnell-Smith, G.P. and Tryon, H. (1923).** The Banana Bunchy Top disease. Qd. Agric. J., **19**. 32-33.

67. **Desai, M.K. (1963).** Checking chlorosis in banana plants. Indian Hort., **7**: 9-13.

68. **Desai, M.V. and Bhide, V.P. (1956).** Fruits and its diseases. The Farmer, June 1956.

69. **Daniells, J.W., Geering, A.D.W. and Thomas, J.E. (1998).** Banana streak virus investigations in Australia. Infomusa. **7**: 20-21.

70. **Daniells, J.W., Geering, A.D.W., Bryde, N.J. and Thomas, J.E. (2001).** The effect of banana streak virus on the growth and yield of dessert bananas in tropical Australia. Ann. Appl. Biol. **139**: 51-60.

71. **Devergne, J.C., Cardin, L., Burekard, J. and Van Regenmortel, M.H.V. (1981).** Comparison of direct and indirect ELISA for detecting antigenically related Cucumovirus. Jour. of Virological Methods. **3**: 193-200.

72. **Diekmann, D.R. and Putter, C.A.J. (1996).** *Musa, 2nd edn. FAO/ I /PGR/Technical Guidelines for the safe movement of Germplasm*, No. 15, FAO Rome IIPGRI Rome, 28 pp.

73. **Dietzgen, R.G. and Thomas, J.E. (1991).** Properties of virus-like particles associated with banana bunchy top disease in Hawaii, Indonesia and Tonga. Australian Plant Path., **20**: 161-165.

74. **Douine, L., Quiot, J.B., Marchoux. G., Archange, P. (1979).** (Revised list of plant species susceptible to Cucumber Mosaic virus (CMV). Bibliographic study). Annales de Phytopathologie **11**: 439-475.

75. **Drew, R.A., Moisander, J.A. and Smith, M.K. (1989).** The transmission of banana bunchy top virus in micro propagated bananas. Plant Cell, Tissue and Organ Culture **16**: 187-193.

76. **Eastwood, H.W. (1933).** Bunchy top control. Early identification, eradication of infective aphids and destruction of diseased stools. Agric. Gaz. N.S.W., **44**: 611-614.

77. **Eastwood, H.W. (1946).** Bunchy top disease of bananas controlled by co-operative effort. Agric. Gaz. N.S.W., **57**: 571-577.

78. **Eastwood, H.W. (1947).** Bunchy top disease of bananas controlled by co-operative effort. Agric. Gaz. N.S.W., **58**: 26-30.

79. **Edwards, M.C. and Gonsalves, D. (1983).** Grouping of seven biologically defined isolates of cucumber mosaic virus by peptide mapping. Phytopathology **73**: 1117-1120.

80. **Eiras, M., Colariccio, A. and Chavies, A.L.R. (2001).** Isolate of cucumber mosaic virus obtained from banana in Sao Paulo State, Brazil, belongs to the subgroup Ia. Fitopatologia Brasileria **26**: 53-59.

81. **El-Deeb, S., Nahla, I., El-Amrity, A. and El-Abbas, F.A. (1997).** Effect of cucumber mosaic virus on the ultrastructure of banana mesophyll cells. Annals of Agric. Sci. (Cairo), **42**: 597-605.

82. **Espino, T.M., Exconde, S.B., Zipagan, F.B. and Espino, R.R.C. (1990).** Banana bract mosaic, a new disease of banana II. Isolation and purification for monoclonal antibody production. The Philippine Agriculturist **73**: 61-68.

83. **Estelitta, S., Radhakrishnan, T.C. and Paul, J.S. (1996).** Infectious chlorosis disease of banana in Kerala. Infomusa, **5**: 25-26.

84. **Evers, C. (1968).** Host plants of sixteen aphids from plantation in Honduras. Flor. Ent. Soc., **1**: 113-118.

85. **Fahmy, T. (1924).** Banana disease caused by a species of *Heterodera.* Egypt, Ministry of Agric. Bull. No. 30.

86. **Feakin, S.D. (1972).** *Pest control in banana,* London: Foreign office Overseas Development Administration, 1972, 128 P, PANS, manual No.1.

87. **Fireman, I.D. (1970).** Crop protection problems of banana in Fiji, PANS, **16**: 625-631.

88. **Foure, E. and Manser, P.D. (1982).** [Note on the appearance in Gabon of a serious virus disease of banana and plantain: bunchy top]. Fruits **37**: 409-414.

89. **Francki, R.I.B., Mossop, D.W. and Hatta, T. (1979).** Cucumber mosaic virus. CMI/AAB. Descriptions of Plant Viruses, No. 213 (No.1 Revised).

90. **Froggatt, W.W. (1923).** The banana aphid (*Pentalonia nigronervosa* Coq.). Agri. Gaz. N.S.W., **34**: 296-297.

91. **Gadd, C.H. (1926).** "Bunchy top" disease of plantain. Trop. Agriculturist **86**: 320.

92. **Gambley, C.F. and Thomas, J.E. (2001).** Molecular characterization of banana mild mosaic virus, a filamentous virus in *Musa* spp. Archives of Virology **146**: 1369-1379.

93. **Gauhl, F. and Pasberg—Gauhl, C. (1994).** Symptoms associated with banana streak virus (BSV). Plant Health Management Division, International Institute of Tropical Agric, Ibadan, Nigeria, 20 pp.

94. **Gauhl, F. and Pasberg—Gauhl, C. (1997).** First report of banana streak badnavirus in plantain landraces in southern Cameroon, Central Africa. Plant Disease **81**: 1335.

95. **Gauhl, F., Pasberg—Gauhl, C., Bopda—Waffo, A., Hughes, J.d'A. and Chen, J.S. (1999a).** Occurrence of banana streak badnavirus on plantain and banana in 45 villages in Southern Cameroon, Central Africa. Zeitchrift fur Pflanzen Krankheiten and Pflanzen Schutz **106**: 174-180.

96. **Gauhl, F., Pasberg—Gauhl, C., Lockhart, B.E.L., Hughes, J.d'A., and Dahal, G. (1999b).** Incidence and distribution of banana streak badnavirus in the plantain production region of South Nigeria. International Jour. of Pest Management **44**: 167-171.

97. **Geering, A.D.W. and Thomas, J.E. (1996).** A comparison of 4 serological tests for the detection of BBTV in banana. Australian Jour. of Agric. Res., **47**: 403-412.

98. **Gibbs, A.J., and Harrison, B.D. (1970).** Cucumber mosaic virus. Description of plant viruses. CMI/AAB, No.1.

99. **Goddard, E.J. (1925).** "Bunchy top" in bananas. Queensland Agric. J., **24**: 424- 429.

100. **Goddard, E.J. (1929).** Bunchy top in bananas. J. Austr. Council Sci. and indust. Res., **2**: 21-27.

101. **Gold, A.H. (1972).** Seed transmission of banana viruses. Phytopathology **62**: 760 (abstr.).

102. **Gomathi, N., Mariappan, V. and Devanathan, M. (1993).** Changes in oxidative enzymes/phenylalanine ammonialyse in mosaic and bract mosaic virus infected banana plants. South Indian Hort., **41**: 376-378.

103. **Govindaswamy, C.V., Padmanabhan, P. and Alagianagalingam, M.N. (1977).** Studies on bunchy top disease of banana. Madras Agric. J., **64**: 205-206.

104. **Graham, K.M. and Navaratnam, S.J. (1970).** Present status of banana diseases in Asia and the Far East. Inf. Lett., FAO Pl. Prot. Comm., S.E. Asia Pacific region 72, 15 pp.

105. **Gupta, P.P. (1986).** Eradication of mosaic disease and rapid clonal multiplication of bananas and plantains through meristem tip culture. Plant Cell, Tissue and Organ Culture **6**: 33-39.

106. **Hafner, G.J., Harding, R.M. and Dale, J.E. (1997).** A DNA primer associated with banana bunchy top virus. Jour. Gen. Virol., **78**: 479-486.

107. **Harding, R.M., Burns, T.M. and Dale, J.L. (1991).** Virus-like particles associated with banana bunchy top disease contain single stranded DNA. Jour. Gen. Virol., **72**: 225-230.

108. **Harding, R.M., Burns, T.M., Hafner, G.J., Dietzgen, R.G. and Dale, J.L. (1993).** Nucleotide sequence of one component of banana bunchy top virus genome contains the putative replicate gene. Jour. of Gen. Virol., **74**: 323-328.

109. **Harper, G. and Hull, R. (1998).** Cloning and sequence analysis of banana streak virus DNA. Virus Genes **17**: 271-278.

110. **Harper, G., Dahal, G., Thottapilly, G., and Hull, R. (1999).** Detection of episomal banana streak badnavirus by IC-PCR. Jour. Virological Methods **79**: 1-8.

111. **Harrison, B.D., Finch, J.T., Gibbs, A.J., Hollings, M., Shepherd, R.J., Valenta, V. and Wetter, C. (1971).** Sixteen groups of plant viruses. Virology **45**: 356-363.

112. **Hassan, M.S. (1958).** Testing phosphorus compounds and chlorinated hydrocarbons against aphids. Agri. Res. Rev., (Engl. edn) **36**: 79-93.

113. **Hasse, A., Ritcher, J. and Rabenstein, F. (1989).** Monoclonal antibodies for detection and serotyping of cucumber mosaic virus. J. Phytopath., **127**: 129-136

114. **He ZiFu, Li Hua Ping, Xiao Huo Gen and Fan Huai Zhong (2000).** Cloning and sequencing of DNA component 6 of BBTV. Chinese. J. of Virology **16**: 185-188.

115. **Hector, G.P. (1925).** Ann. Rept. Economic Botanist to Govt. of Bengal for the year 1923-24.

116. **Hu, J.S., Li, H.P., Barry, K., Wang, M. and Jorden, R. (1995).** Comparison of dot blot, ELISA, and RT-PCR assay for detection of two cucumber mosaic virus isolates infecting banana in Hawaii. Plant Dis., **79**: 902-906.

117. **Hu, J.S., Wang, M., Sether, D., Xie, W. and Leonhardt, K.W. (1996).** Use of polymerase chain reaction (PCR) to study transmission of banana bunchy top virus by the banana aphid (*Pentalonia nigronervosa*). Ann. Appl. Biol. **128**: 55-64.

118. **Hughes, J.d'A., Speijer, P.R. and Olatunde, O. (1998).** Banana Die back virus a new virus infecting banana in Nigeria. Plant Disease **82**: 129.

119. **Hwang, S.C. (1991).** Status of banana diseases in Taiwan. In: *Banana diseases in Asia and Pacific.* Eds. Valmayor, R.V., Umali, E., Bejosano (Vistina) P. INIBAP Network for Asia and the Pacific, p. 73-83.

120. **Iskra, M.L. and Bove, J.M. (1986).** Virus diseases of banana: Methods for early detection and characterization of bunchy top associated virus. In: *Seminar on improving citrus and banana production in the Caribbean through Phytosanitation.* St. Luna (LCA) 1986, CTA Wageningen (NLD) (1987): 29-32.

121. **Iskra, M.L., Garnier, M. and Bove, J.M. (1989).** Purification of banana bunchy top virus. Fruits **44**: 63-66.

122. **James, M., Kenten, R.H., and Woods, R.D. (1973).** Virus like particles associated with two diseases of *Colocasia esculenta* (L.) Schott in the Soloman Islands. J. Gen. Virol., **21**: 145-153.

123. **Jayaraj. S. (1987).** Recent studies on plant protection in banana in Tamil Nadu. Paper presented at the All India Coordinated Fruit Improvement Project Workshop held at Nagpur on July 27-30, 1987.

124. **Johnson, J. (1963).** Biology of the bunchy top aphid *Pentalonia nigronervosa* Coq. (Aphididae: Hemiptera). Agric. Res. J. Kerala **2**: 45-51.

125. **Johnson, J. (1965).** Relative toxicity of some synthetic organic insecticides to the banana aphid, *Pentalonia nigronervosa* Coq. Agri. Res. J. Kerala **3**: 55-57.

126. **Johnston, A. (1964).** "Plant disease in Samoa". Pl. Prot. Bull., F.A.O.**12**: 79-81.

127. **Jones, D.R. (ed.)(1995).** Black sigatoka disease inVenezuela. Infomusa, 4: 13-14 Jones, D.R. (ed.) Diseases of Banana, Abaca and Enset. CAB Publishing, New York, 544 pp.

128. **Jones, D.R., and Lockhart, B.E.L. (1993).** *Banana streak disease. Musa* Disease Fact Sheet No.1. International Network for the Improvement of Banana and Plantain, Montpellier, France.

129. **Jose, P.C. (1981).** Reaction of different varieties of banana against bunchy top disease. Agri. Res. J. Kerala **19**: 108-110.

130. **Jose, P.C., Balagopal, C. Wilson, K.I. and Nambiar, E.P. (1971).** Effect of bunchy top virus infection on the chemical constituents of banana fruits. Agri. Res. J. Kerala **9**: 96-97.

131. **Joshi, D.M. and Joshi, H.U. (1974).** The control of banana Mosaic. Indian J. Hort., **31**: 382-385.

132. **Joshi, D.M. and Joshi, H.U. (1976).** Occurrence of banana Mosaic in South Gujarat area and its host range. Guj. Res. J.,**1**: 105-110.

133. **Kadhirvel, A.K., Sathiamoorthy, S., Baskaran, T.L. and Letchoumanane, S. (1986).** Outbreak of banana mosaic in Trichi District. TNAU News Letter **15**: 3.

134. **Kagy, V., Thomas, J.E., Sharman, M. and Mademba–Sy, F. (2001).** First record of banana bunchy top disease in New Caledonia. Australasian Plant Pathology **30**: 71.

135. **Kamat, M.N. and Patel, M.K. (1951).** Notes on two important plant diseases in Bombay. Stat. Bull. Pl. Prot., New Delhi, **3**: 16.

136. **Kao, C.W. and Faan, H.C. (1983).** [On a CMV strain causing the banana mosaic in Guangzhou.] Journal of South China Agricultural College **4**: 43-48.

137. **Karan, M., Harding, R.M. and Dale, J.L. (1994).** Evidence for two groups of bunchy top virus isolates. Jour. Gen. Virol., **75**: 3541-3546.

138. **Kawano, S. and Su, H.-J. (1993).** Occurrence of banana bunchy top virus in Okinawa. Annals of the Phytopath. Soc. of Japan **59**: 53.

139. **Kenyon, L., Magnaye, L., Warburton, H., Herradura, L., Chancellor, T., Escobido, E. and Foot, C. (1996).** *Epidemiology and Control of Banana Virus Diseases in the Philippines.* NRI—Dept. for International Development Crop Prot: Program Project A0217/X0258 Final Techn. Report, National Resource Institute, Chatham Mantime, U.K.

140. **Kenyon, L., Brown, M. and Khonje, P. (1997).** First report of banana bunchy top virus in Malawi. Plant Disease **81**: 1096.

141. **Keshavamurthy, R.C. (1980).** Radical changes in ecosystem in the Palani Hills. In: Proc. Natl. Seminar on Banana Production Technol. (eds. Muthikrishnan, C.R. and Abdul Khader, J.B.M. Md.) TNAU., Coimbatore, pp. 23-28.

142. **Khalid, S. and Soomro, M.H. (1993).** Banana bunchy top disease in Pakistan. Plant Path., **42**: 923-926.

143. **Khalid, S., Soomro, M.H. and Stover, R.H. (1993).** First report of banana bunchy top virus in Pakistan. Plant Disease **77**: 101.

144. **Khurana, S.M.P. (1971).** Occurrence of the banana aphid and its relation with banana bunchy top disease in Gorakhpur (U.P.). Indian J. Hort., **28**: 167-168.

145. **Kiranmai, G., Satyanarayana, T. and Sreenivasulu, P. (1997).** Molecular cloning and detection of cucumber mosaic cucumovirus causing infectious chlorosis of banana using DNA probes. Indian Phytopath. Soc. Golden Jublee Int. Conf., Nov. 10-15, 1997. New Delhi.

146. **Kolkaila, A.M. and Soliman, A.A. (1954).** A study of the banana aphid, *Pentalonia nigronervosa* Coq. (Hemiptera, Homoptera, Aphididae). Bull. Soc., *Fouadier lat. Le* Caire **38**: 231-250.

147. **Kubiriba, J., Legg, J.P., Tushemereirwe, W. and Adipala, E.**

(2001). Vector transmission of banana streak virus in the screen-house in Uganda. Ann. Appl. Biol., **139**: 37-43.

148. **Kubiriba, J., Tushemereirwe, W. and Karamura, E.B. (1977).** Distribution of Banana Streak Virus (BSV) in Uganda. Mus Africa **11**: 17.

149. **La Fleur, D.A., Lockhart, B.E.L. and Olszewski, N.E. (1996).** Portion of banana streak badnavirus genome are integrated in the genome of its host, *Musa*. Phytopathology **86**: 100.

150. **La Ping, Cai Ven Qi and Fan Rong Xiang.** [Cloning and sequencing of banana bunchy top virus genome component 1]. Chinese Jour. Of Virology **16**: 158-161.

151. **Lassoudiere, A. (1974).** La mosaique dite "a tirets" due bananier 'Poyo' en Cote d'lvoire. Fruits **29**: 349-357.

152. **Lassoudiere, A. (1979).** Account of the economic repercussions of banana dotted line mosaic in the Ivory Coast. Possibility of control by eradication. Fruits **34**: 3-34, 77-80.

153. **Lockhart, B.E.L. (1986).** Purification and serology of a bacilliform virus associated with banana streak disease. Phytopathology **76**: 995-999.

154. **Lockhart, B.E.L. (1990).** Evidence for a double stranded circular DNA genome in a second group of plant viruses. Phytopathology **80**: 127-131.

155. **Lockhart, B.E.L. (1994a).** Banana streak virus. In: Ploetz, R.C., Gentmeyer, G.A., Nishijima, W.T., Rohrbach, K.G. and Ohr, H.D. (eds). *Compendium of Tropical Fruit Diseases*. Amer. Phytopath Soc. Press, St Paul, Minnesota, USA, 19-20.

156. **Lockhart, B.E.L. (1994b).** Development of detection methods for banana streak virus. In: *The Global Banana and Plantain Network*, INIBAP Annual Report 1994, INIBAP, Montpellier, France, pp. 20-21.

157. **Lockhart, B.E.L. (1995).** Banana streak badnavirus infection in *Musa*: Epidemiology, diagnosis and control, ASPAC Food

and Fertilizer Technology Center, Taiwan, Techn. Bull.,**143**: 11 pp.

158. **Lockhart, B.E.L. and Autrey, L.J.C. (1988).** Occurrence in sugarcane of a bacilliform virus related serologically to banana streak virus. Plant Disease **72**: 230-233.

159. **Lockhart, B.E.L. and Olszewski, N.E. (1993).** Serological and genomic heterogeneity of banana streak badnavirus. Implications for virus detection in *Musa* germplasm. In: J. Ganry (ed), Breeding Banana and Plantain for Resistance to Diseases and Pests. CIRAD/IRFA/INIBAP. Montpellier, France, 393 pp.

160. **Lockhart, B.E.L., Autrey, L.J.C. and Comstock J.C. (1992).** Partial purification and serology of sugarcane mild mosaic virus, a mealybug transmitted clostero like virus. Phytopathology **82**: 691-695.

161. **Lockhart, B.E.L., Bouhida, M. and Olszewksi, N.E. (1988).** Evidence for a double stranded DNA genome in small non-enveloped bacilliform plant viruses. Phytopathology **78**: 1959.

162. **Lockhart, B.E.L. Ndowora, T.C., Olszewksi, N.E. and Dahal, G. (1988a).** Studies on integration of banana streak badnavirus sequences in *Musa:* identification of episomally-expressible badnaviral integrants in *Musa* genotype. In: E.A. Frison and S.L. Sharrock eds. Proc. Workshop of the PRO MUSA Virology working group, Montpellier, France, 1921 Jan. 1996, pp. 42-47.

163. **Lockhart, B.E.L., Ndowora, T.C., Olszewksi, N.E. and Dahal, G. (1988b).** Viral leaf streak of *Musa*: investigating a novel virus host interaction. In: International Seminar on Plantain Production, Armenia, Quindio, Columbia, May, 1998.

164. **Lockhart, B.E.L., Ndowora, T.C., Olszewski, N.E. and Dahal, G. (1998).** Studies on integration of banana streak badnavirus sequences in *Musa*: identification of episomally-expressible badnaviral integrants in *Musa*

genotypes. Banana streak virus: a unique virus-*Musa* interaction. Proc. of a workshop of the PROMUSA Virology working group Montpellier, France, Jan. 19-21, 1998.

165. **Magee, C.J.P. (1927).** Investigation on the bunchy top disease of banana. Comm. Austr. Countries Sci. Ind. Res. Bull., 30, p. 64.

166. **Magee, C.J.P. (1930).** A new virus disease of banana. Agric. Gaz. N.S.W., **41**: 929.

167. **Magee, C.J.P. (1936).** Bunchy top disease of banana. Rehabilitation of the banana industry in New South Wales. J. Austr. Inst. Agri. Sci., **2**: 13-16.

168. **Magee, C.J.P. (1939).** Pathological changes in the phloem and neighbouring tissues of the banana (*Musa Cavendish* Lamb.) caused by bunchy top virus. Sci. Bull. Dept. Agri. N.S.W., **67**: 32.

169. **Magee, C.J.P. (1940a).** Transmission studies on the banana bunchy top virus. J. Aust. Inst. Agri. Sci., **6**: 109-110.

170. **Magee, C.J.P. (1940b).** Transmission of infectious chlorosis or heart rot of the banana and its relationship to cucumber mosaic. J. Austr. Inst. Agric. Sci., **6**: 109-110.

171. **Magee, C.J.P. (1948).** Transmission of bunchy top of banana varieties. J. Aust. Inst. Agric. Sci., **14**: 18-24.

172. **Magee, C.J.P. (1953).** Some aspects of bunchy top disease of banana and other *Musa* spp. J. Proc. R. Soc. N.S.W., **87**: 1-18.

173. **Magee, C.J.P. (1967).** The control of banana bunchy top. New Caledonia, South Pacific Commission,1967, SPC Tech. Paper 150.

174. **Magnaye, L.V. and Eloja, A.L. (1968).** Banana mosaic in the Philippines: I. Transmission and initial host range studies. 5th Annual Conf. on the Philippine Phytopath. Society, May 10-13,1968, Davao city.

175. **Magnaye, L.V. and Espino, R.R.C. (1990).** Note: Banana bracts mosaic a new disease of banana. I. Symptomatology. The Philippine Agriculturist **73**: 55-59.

176. **Magnaye, L.V. and Valmayor, R.V. (1995).** BBTV, CMV and other viruses affecting banana in Asia and South Africa. African Plant Prot., **2**: 1-8.

177. **Mali, V.R. and Deshpande, G.D. (1976).** Heart rot- a virus disease of banana in Marathwada. Indian J. Mycol. Pl. Path., **6**: 23-26.

178. **Mali, V.R. and Rajegore, S.B. (1980).** A cucumber mosaic virus disease of banana in India. Phytopath. Z., **98**: 127-136.

179. **Manickam, K., Sabitha Doraiswamy, Ganapathy, T., Mala, G.T. and Rabindran, R. (2001).** Characterization and serological detection of banana bunchy top virus in India. Z. Pfl Krankh Pfl Schutz., **108**: 490-499.

180. **Manser, P.D. (1982).** Bunchy top disease of plantain. F.A.O. Pl. Prot. Bull., **30(2)**: 78-79.

181. **Marathe, T.S., Summanwar, A.S. and Ram, R.D. (1982).** New additional hosts of *Pentalonia nigronervosa* Coq. Curr. Sci., **51**: 435-436.

182. **Mariappan, V. and Mathikumar, P. (1992).** Serological detection of bunchy top virus disease of banana. South Indian Hort., **40**: 231-232.

183. **Matthews, R.E.F. (1982).** Classification and nomenclature of viruses. Intervirology **17**: 1-200.

184. **Matile-Ferroro, D. and Wiliams, D.J. (1995).** Recent outbreaks of mealy bugs on plantain (*Musa* spp.) in Nigeria including a new record for Africa and a description of a new species of *Planococcus ferris* (Homoptera, Pseudocociidae). Bulletin de la Societe Entomologique de France**100**: 445-449.

185. **May, G.M., Afza, R., Mason, H.A., Wiccko, A., Novak, F.J. and Arntzen, C.J. (1995).** Generation of transgenic banana (*Musa acuminata*) plants via Agrobacterium-mediated transformation. Bio/Technology **13**: 486-492.

186. **Mayo, M.A. (2002).** The principles and current practice of plant virus taxonomy In : Plant Viruses as Molecular

Pathogens (Jawaid A. Khan and Jeanne Dijkstra eds.) Chapter 1, pp. 3-23.

187. **Medberry, S. L., Lockhart, B. E. L. and Olszewski, N. E. (1990).** Properties of *Commelina* yellow mottle virus complete transcript suggests that it is a pararetrovirus. Nucleic Acid Res., **18**: 5055-5513.

188. **Mehta, P.R., Joshi, N.C., Rao, M.H. and Renjhen, P.L. (1964).** Bunchy top- serious disease of banana in India. Sci & Cult., **30**: 259-263.

189. **Menon, M.R. and Christudas, S.P. (1966).** Investigation on the control of the banana aphid, *Pentalonia nigronervosa* Coq. Agri. Res. J. Kerala **4**: 92-94.

190. **Menon, M.R. and Christudas, S.P. (1967).** Studies on the population of the aphid *Pentalonia nigronervosa* Coq. on banana plants in Kerala Agric Res. J. Kerala **5**: 84-86.

191. **Meredith, D.S. (1970).** Major banana diseases. Past and present status. Rev. Pl. Path., **49**: 539-554.

192. **Mohan, M.K. and Rao, V.N.M. (1986).** Distribution of P32 in healthy and bunchy top diseased 'Robusta' banana. Indian J. Agric. Sci., **56**: 543 - 544.

193. **Mohan, S. and Lakshmanan, P. (1988).** Outbreak of cucumber mosaic virus on *Musa* sp. in Tamil Nadu, India. Phytoparasitica **16**: 281-282.

194. **Morgan, C.N. (1969).** Banana bunchy top control in Queensland. Paper presented to pacific-Asian Banana Culture Interchange, Australia, May, 1969.

195. **Munez, A.R. (1992).** Symptomatology, transmission and purification of banana bract mosaic virus (BBMV) in 'Giant Cavendish' banana. M.Sc. Thesis, University of the Philippines, Los Banos, Philippines.

196. **Murant, A.F., Waterhouse, P.M., Rashke, J.H. and Robinson, D.J. (1985).** Carrot red leaf and carrot mottle viruses: observations on the composition of the particles in single and mixed infections. Jour. Gen. Virol., **66**: 1575-1579.

197. **Nair, C.K.N. and George, C.K. (1966).** Nutritional status of soils and the incidence of 'bunchy top' disease of banana (*Musa* sp). III Calcium oxide/Magnesium oxide ratio in plants as a function of the soil contents of calcium and magnesium and its relationship to 'Bunchy top' infection. Agri. Res. Jour. Kerala **4**: 106-121.

198. **Nair, C.K.N. and Pillai, K.S. (1966).** Nutritional status of soil and the incidence of 'bunchy top' disease of banana (*Musa* sp.). II. Significance of calcium/magnesium ratio in the nutrient medium and in plants. Agric. Res. J. Kerala **4**: 86-105.

199. **Nair, C.K.N., Srinivasan, K. and Amma, J.S. (1966).** Nutritional status of soils and the incidence of the 'Bunchy top' disease of the bananas (*Musa* sp.) IV. Anatomical variations in virus infected and healthy plants as a function of calcium/magnesium ratio in soils. Agric. Res. Jour. Kerala **4**: 122-136.

200. **Nair, M.R.G.K., Jose, P.C., Regunath, P. and Nair, N.G. (1973).** Effect of some insecticide granules on the control of the banana aphid, *Pentalonia nigronervosa* Coq. Agri. Res. Jour. Kerala **11**: 101-102.

201. **Nair, T.C. and Menon, M.R. (1965).** Beating back the bunchy top. Indian Fmg., **15**: 23.

202. **Nair, P.K.B. and Wilson, K.I. (1970).** Effect of bunchy top virus infection on the free amino acid and amide composition of banana leaves. Agri. Res. Jour. Kerala **8**: 137-138.

203. **Nair, P.K.B. and Wilson, K.I. (1972).** Effect of bunchy top virus infection on the carbohydrate and nitrogen contents of banana leaves. Agri. Res. J. Kerala **10**: 38-41.

204. **Nambiar, P.K.D. and Nair, C.K.N. (1965).** Investigations on the possible relationship between the nutritional status of the soil and the incidence of 'bunchy top' disease of banana (*Musa* sp.). Agric. Res. Jour. Kerala **3**: 78-79.

205. **Nandi, H.K. (1941).** Rept, Dept. Agric. Assam, (1939-40): 85-138.

206. **Nayar, G. (1962).** Banana in India. Udyogamandal. The Fact Technical Society.

207. **Nayar, R. (1984).** Association of nematodes in bunchy top of banana. Curr. Sci., **53**: 595-597.

208. **Ndowora, T.C., Lockhart, B.E.L. and Olszewski, N.E. (1997).** Relationship between integrated and episomal badnavirus genomic sequences in *Musa*. Phytopathology **87**: 569.

209. **Ndowora, T.C. (1998).** Banana streak virus : development of a immunoenzymatic assay for detection and characterization of sequences that are integrated in the genome of the nost. Ph.D. Thesis, University of Minnesota, U.S.A. 90 pp.

210. **Ndowora, T., Dahal, G., La Fleur, D., Harper, G., Hull, R., Olszewski, N.E. and Lockhart, B.E.L. (1999).** Evidence that badnavirus infection in *Musa* can originate from integrated pararetroviral sequences. Virology **255**: 214-220.

211. **Niblett, C.L., Pappu, S.S., Bird, J. and Lastra, R. (1994).** Infectious chlorosis, mosaic and heart rot. In: Ploetz, R.C., Zentmyer, G.A., Nishijima, W.T., Rohrbach, K.G. and Ohr., H.D. (eds). Compendium of Trop. Fruit Diseases. APS Press, St Paul, Minnesota, pp.18-19.

212. **Nitzany, F.E. (1975).** Cucumber mosaic virus in Israel. Phytopath. Medit.,**14**:16-20.

213. **Ocfemia, G.O. (1926).** Progress report on bunchy top abaca or Manila hemp. Phytopathology **16**: 894.

214. **Ocfemia, G.O. (1930).** Bunchy top of Abaca and Manila Hemp. I. A study of the cause of the disease and its method of transmission. Amer. J. Bot., **17**: 1-18.

215. **Ocfemia, G.O. and Buhay, G.G. (1934).** Bunchy top of Abaca or Manila Hemp. II. Further studies on the transmission of the disease and a trial planting of abaca seedlings in a bunchy top devastated field. Philipp. Agriculturist **22**: 567-581.

216. **Ocfemia, G.O., Celini, S. and Garcia, J. (1947).** Further studies on transmission of bunchy top and mosaic of Abaca (Manila Hemp plant). Separation of the two diseases and mechanics of inoculation by *Pentalonia nigronervosa* Coq. Philipp. Agriculturist **31**: 87-97.

217. **Ogilvic, L. (1927).** Agri. Bull. Bermuda, Dept. Agric., **6**: 57 pp.

218. **Omura, T., Saito, Y., Usugi, T., and Hibino, H. (1983).** Purification and serology of rice tungro spherical and rice tungro bacilliform viruses. Ann. Phytopathol Soc. Japan **49**: 73-76.

219. **Osei, J.K. (1995).** Cucumber mosaic virus (CMV) infects plantain in Ghana. Mus Africa, No. **6**: 4-5.

220. **Othman, B.A., El-Dougdoug K.A., Sadik, A.S. and Aabdel - Ghaffar, M.H. (1996).** Detection of banana bunchy top virus in some banana plantations in Kalubia Governorate. Annals Agric Sec. (Cairo), **41**: 627-634.

221. **Owen, J. and Palukaitis, P. (1988).** Characterization of cucumber mosaic virus. Virology **166**: 495-502.

222. **Padwick, G.W. (1940).** Sci. Rept. Agric. Res. Inst. New Delhi (1938-39).

223. **Palukaitis, P., Roossinck, M.J., Dietzgen, R.G. and Francki, R.l.B. (1992).** Cucumber mosaic virus. Adv. Virus Res., **41**: 281-348.

224. **Pappu, S.S., Pappu, H.R., Niblett, C.L. and Bird, J. (1991).** Molecular cloning and nucleotide sequencing of isolates of cucumber mosaic virus from bananas and plantains in Puerto Rico (Abstr.). Phytopathology **81**: 1154.

225. **Pasberg-Gauhl, C., Gauhl, F. and Schill, P. (1995).** First report of banana streak virus in Farmers fields in Benin, Ghana and Nigeria, West Africa. Plant Disease **80**: 224.

226. **Patil, K.V. (1982).** Further studies on cucumber mosaic virus causing banana mosaic or heart rot disease. Ph.D. Thesis Marathwada Agric. Univ. Parbhani, 104 pp.

227. **Pietersen, G., Stapes, M.S., Kasdorf, G.G.F. and Thomas, J.E. (1998).** First report of cucumber mosaic cucumovirus subgroup 1 in South Africa, from banana with infectious chlorosis. Plant Disease **82**: 1171.

228. **Pillai, P.B. and Nair, C.K.N. (1974).** Nutritional status of soils and the incidence of bunchy top disease of banana (Var. Java). Agric. Res. Jour. Kerala **12**: 135-139.

229. **Ploetz, R.C. (1994).** Bract mosaic. In: Ploetz, R.C., Zentmyer, G.A., Nishijima, W.T., Rohrbach, K.G. and Ohr, H.D. (eds.). Compendium of Tropical Fruit Diseases. APS Press, Minnesota.

230. **Ram, R.D. and Singh, S.J. (1999).** Occurrence of banana streak badnavirus in Maharashtra. In: National Seminar on "Recent Advances in Plant Pathology", Dept. of Botany, University of Pune, Jan. 7-8, 1999.

231. **Ram, R.D. and Summanwar, A.S. (1984).** *Colocasia esculenta* (L) Schott, a reservoir of bunchy top disease of banana. Curr. Sci., **53**: 145-146.

232. **Ramaswamy, S. (1967).** Studies on bunchy top disease of banana. M.Sc. (Ag.) Dissert. Univ. Madras.

233. **Ramaswamy, S., Ramakrishnaiah, K. and Govindaswamy, C.V. (1966).** A note on the effect of gibberellic acid on bunchy top disease of banana. Madras Agric. J., **53**: 513-515.

234. **Rao, D.G. (1980a).** A short note on bunchy top disease of banana. Proc. National Seminar on Banana Production Technol. (Abstr). p. 26.

235. **Rao, D.G. (1980b).** Studies on a new strain of banana mosaic virus in South India. In: *Proc. of Natl. Seminar on Banana Production Technology* (edt.). Muthukrishnan, C.R. and Abdul Khader, J.B.M. Md.) TNAU., Coimbatore, P. 155-159.

236. **Reddy, M.K., Vani, A., Selvarajan, R. and Singh, H.P. (1996).** Immunological and molecular diagnosis of banana bract mosaic virus (Abstr.). In : Conference on Challenges for

Banana Production and utilization, p. 58, 24-25 Sept., 1996, NRC Banana, Trichy.

237. **Regupathy, A. and Kulasekaran, M. (1980).** Studies on the possibilities of bunchy top disease containment through vector control. In: *Proc. Natl. Seminar on Banana Production Technology* (eds.) Muthukrishnan, C.R. and Abdul Khader, J.B.M. Mohd., T.N.A.U., Coimbatore, pp. 150-154.

238. **Regupathy, A. and Subramanian, K.S. (1980).** Studies on the combined use of insecticide and weedicide in banana bunchy top disease management program. Proc. Seminar and Diseases and Manuring of Plantation Crops, Madurai, pp. 6-8.

239. **Regupathy, A., Subramanian, K.S. and Naganathan, T.G. (1983).** Evaluation of certain aphidicides in the containment of banana bunchy top. Pesticides **17**: 35-36.

240. **Reichel, H., Belalcazar, S., Munera, G., Arevalo, E. and Narvaez, J. (1996).** First report of banana streak virus infecting plantains (*Musa* spp.) in Columbia. Plant Dis., **80**: 463.

241. **Reichel, H., Belacazar, S., Munera, G., Perez, R. and Arevalo E. (1997).** The presence of banana streak virus has been confirmed in plantain (*Musa* AAB Simmonds), sugarcane (*Saccharum officinarum*) and edible canna (*Canna dulis*) in Colombia. Infomusa **6**: 9-12.

242. **Reinking, O.A. (1950).** Banana virus studies and observations made at Taiwan, Sempora, Mostyn and Sebatik Island, British North Borneo: Suggested studies for investigations to clarify the Abaca bunchy top relationship with the banana virus complex. Pl. Dis. Reptr., **34**: 66-69.

243. **Retuerma, M.L. (1981).** Mosaic of abaca and banana : Their causal viruses and properties. Ph.D. Dissertation. Univ. of Philippines at Los Banos, College, Laguna.

244. **Reynolds, P.K. (1951).** Earliest evidence of banana culture. J. Amer. Orient. Soc. Suppl., **12**: 28.

245. **Richardson, D.L., Hamilton, K.S. and Hutehison, D.J. (1965).** Notes on bananas, I, Natural edible tetraploids. Trop. Agric. (Trinidad), **42:** 125-137.

246. **Rivas, S. (1988).** Production of plants of two banana cultivars free from cucumber mosaic virus by culture of apical meristems isolated *in vitro*. Fitopatologia Venezolana **1:** 69-72.

247. **Rivera, R., Ramirez, P. and Pereira, R. (1992).** Preliminary characterization of viruses infecting banana in Costa Rica-Pages 63-68. In: Proc. Workshop Biotechnol. Apll. Banana Plantain improvement. International Network for the Improvement of Banana and Plantain, Montpellier, France.

248. **Rodoni, B.C., Ahlawat, Y.S., Varma, A., Dale, J.L. and Harding, R.M. (1997).** Identification and characterization of banana bract mosaic virus in India. Plant Dis., **81:** 669-672.

249. **Rohde, W., Randles, J.W., Langridge, P. and Hanold, D. (1990).** Nucleotide sequence of a circular single-stranded DNA associated with coconut foliar decay virus. Virology **176:** 648-651.

250. **Roperos, N.I. and Magnaye, L.V. (1991).** Status of Banana Diseases in the Philippines. Proc. of a Regional Technical Meeting on Diseases Affecting Banana and Plantain in Asia and the Pacific. Brisbane, Australia, 15-18 April, 1991 (ed. Ramon V. Valmayer).

251. **Sagi, L., Panis, B., Remy, S., Schoofs, H., De Smet, K., Swennen, R. and Cammue, B.P.A. (1995).** Genetic transformation of banana and plantain (*Musa* spp.) via particle bombardment. Bio/Technology **13:** 481-485.

252. **Sagi, L. Remi, S., Cammue, B.P.A., Maes, K., Raemaebers, T., Panis, B., Schoofs, H. and Swennen, R. (2000).** Production of transgenic banana and plantain. Acta Hort., **540:** 203-206.

253. **Sadik, A.S. and El-Karim, G.G. (1997).** Detection of banana bunchy top nanavirus in banana aphid (*Pentalonia nigronervosa*). Annals of Agric. Sci. (Cairo), **42:** 17-28.

254. **Salih, H. and Nas, Y.Z. (1974).** Preliminary studies on banana mosaic (Cucumber mosaic) virus found on banana (*Musa cavendishii*). J. Turkish Phytopath., **3**: 83-88.

255. **Samraj, J., Menon, M.R. and Christudas, S.P. (1970).** The movement of banana bunchy top virus in plant. Agric. Res. Jour. Kerala **8**: 106-108.

256. **San Juan, M.N. (1985).** Foliar diseases of banana and their control in commercial Plantation. In: Proc. Banana and Plantain Research and Development. ACIR and PCARRD. Los Banos, Laguna (Philippines). PCARRD Book Series No. **41**: 181 pp.

257. **Sastry, K.S., Rao, D.G. and Singh, S.J. (1980).** Studies on control of bunchy top of banana. Proc. Nat. Semi. Banana Prod. Tech. (Abstr.): p. 26.

258. **Sebasigari, K. and Stover, R.H. (1988).** Banana diseases and pests in East Africa. Report of a Survey in November, 1987, INI BAP, Montpellier, France, 15 pp.

259. **Selvarajan, R. and Singh, H.P. (1997).** Occurrence, geographical distribution and electron microscopy of BBMV in India (Abstr.). Intln. Conf. On Integrated Plant Disease Management for Sustainable Agric., Nov. 10-15, IARI, New Delhi, p. 327.

260. **Selvarajan, R., Thangavelu, R., Singh, S. J. and Singh, H.P. (1997).** Geographical distribution, serodiagnosis and yield loss assessment of BSV in India (Abstr.). In: Intln. Conf. On Integrated Plant Disease Management for Sustainable Agric. Nov. 10-15, IARI, New Delhi, p. 223.

261. **Shanmuganathan, N. (1980).** Virus and virus-like disease of plants in the Gilbert Islands. FAO Plant Protection Bull., **28**: 29-38.

262. **Sharma, S.R. (1988).** Banana bunchy top virus- a review. Int. J. Trop. Plant Dis., **6**: 19-41.

263. **Sharma, S.R. (1985).** Annual Report. Indian Institute of Horticultural Research, Bangalore, p. 60.

264. **Sharman, M., Gambley, C.F., Olotco, E.O., Abgona, R.V.J. and Thomas, J.E. (2000).** First record of natural infection of abaca (*Musa textilis*) with banana bract mosaic potyvirus in the Philippines. Australian Plant Path., **29**: 69.

265. **Shepherd, K. and Ferreira, F.R. (1984).** The PNG Biological Foundation's Banana Collection at Laloki, Port Moresby, Papua New Guinea. IBPGR Regional Committee for Southeast Asia News letter **8**: 28- 34.

266. **Shimmonds, J.H. (1966).** Host Index of Plant Diseases in Queensland. Brisbane, Australia, Queensland Department of Primary Industries.

267. **Siddappaji, C. and Reddy, D.N.R.N., (1972).** A note on the occurrence of the aphid, *Pentalonia nigronervosa* from Caladil, Vander Goot, (Aphididae-Homoptera) on Cardamom. Mysore J. Agric. Sci., **6**: 194-195.

268. **Silberschmidt, R. and Nobrega, N.R. (1941).** Sobre Uma doenca de virus da bananeira. Biologico **7**: 216-219.

269. **Simmonds, N.W. (1933).** Report on visit to Samoa. Dept. of Agriculture, Fiji.

270. **Simmonds, N.W. (1959).** Bananas Longmans, Green and Co. Ltd. London, p. 512.

271. **Simmonds, N.W. (1962).** *The evolution of the bananas*. London, Longmans, Green.

272. **Simmonds, N.W. (1966).** Bananas. London, Longmans, Green, 2nd edn., 1966.

273. **Simmonds, N.W.(1982).** Bananas. Longman Group Ltd., U.K. p. 512.

274. **Singh, R.N. (1979).** Bunchy top of banana in Uttar Pradesh. Ind. J. Mycol. Pl. Path., **9**: 253-254.

275. **Singh, S.J. (1988a).** Virus, Viroid and Mycoplasma disease of Fruit crops in India(Abstract). Natn. Seminar on Changing Pest Situation in Current Agric. Scenario of India. IARI, New Delhi, 14-16 June 1988.

276. **Singh, S.J. (1988b).** A new mosaic disease of banana. Curr. Sci., **57**: 1248-1249.

277. **Singh, S.J. (1989).** Banana bunchy top in Nagaland (Unpublished).

278. **Singh, S.J. (1996).** Viral diseases of banana. In: Advances in Diseases of Fruit Crops in India (S.J. Sing, ed.). pp. 1-35, Kalyani Publishers, Ludhiana, India.

279. **Singh, S.J. (1999).** Diseases of banana and their management In: Diseases of Horticultural Crops-Fruits (Eds. L. R. Varma and Sharma). Indus Publishing Co., New Delhi, pp. 451-478.

280. **Singh, S.J. and Raj Verma (2002).** Incidence of virus disease in tissue cultured banana plantations in Maharashtra. In: Global Conf. on Banana and Plantain, 28-31 October 2002, Bangalore, India.

281. **Smith, K.M. (1937).** A textbook of plant virus diseases.

282. **Soomro, M.H., Khalid, S. and Aslam, M. (1992).** Outbreak of banana bunchy top virus in Sindh, Pakistan. FAO Plant Protect. Bull. **40**: 95-99.

283. **Stadler, B. and Voelkl, W. (1991a).** Foraging pattern of two aphid parasitoids, *Lysiphlebus testaceipes* and *Aphidius colemani* on banana. Entomol. Expt. Appl., **58**: 221 -273.

284. **Stadler, B. and Voelkl, W. (1991b).** Introduction of Lysiphlebus *testaccipes* (Cresson) (Hym., Aphididae) into the Kingdom of Tonga, Oceania. In: Ecology and effectiveness of Aphidophaga. E. Niemezyk and A.F.G. Dixon (eds). SPE Acad. Publ., The Hague, Netherlands, pp. 271-273.

285. **Stary, P. (1966).** Integrated control of aphids attacking bananas. Rev. Agric. Subtrop. Trop., **60**: 224-241.

286. **Stary, P. (1975).** *Aphidius Colemani* Viereck: its taxonomy, distribution, and host range (Hymenoptera, Aphididae). Acta Entomol. Bohemosiev., **72**: 156-163.

287. **Stary, P. and Stechman, D.H. (1991).** Niche distribution and dispersal of the banana aphid, *Pentalonia nigronervosa* Coq. (*Homoptera*; Aphididae). Jour. Appl. Entomol., **109**: 457-562.

288. **Stechmann, D.H. and Voelkl, W. (1988).** Introduction of *Lysiphlebus testaceipes* (Cresson)(Hym. Aphididae) into the Kingdom of Tonga, Oceania. In: Ecology and Effectiveness of Aphidophaga, E. Neimezyk and A.F.G. Dixon (eds). SPB Acad. Publ., The Hague, Netherlands, pp. 271-273.

289. **Stover, R.H. (1972).** *Banana, Plantain and Abaca Diseases.* Commonwealth mycological Institute, Kew, Surrey, U.K., 319 pp.

290. **Su, H.J. (1998).** First occurrence of banana streak badnavirus and studies on its vector relationship in Taiwan. In: Proc. of a Workshop of the PROMUSA Virology Working Group, Montpellier, France, 19 - 21 Jan., 1998, pp. 20-25.

291. **Su, H.J., Hung, T.H. and Wu, M.L. (1997).** First report of banana streak virus infecting banana cultivars (*Musa* spp.) in Taiwan. Plant Disease **81**: 550.

292. **Su, H.J. and Wu, R.Y. (1988).** Characterization and monoclonal antibodies of the virus causing banana bunchy top (Abstr.). Fifth Int. Congr. Pl. Path., Kyoto, Japan.

293. **Su, H.J. and Wu, R.Y. (1989).** Characterization and monoclonal antibodies of the virus causing banana bunchy top. Technical Bull. No. 115, 10 p. Food and Fertilizer Technology Center, National Univ., Taipei, Taiwan.

294. **Sulyo, Y., Duriat Ati, S. and Said dan, A. (1978).** Peninjauan pendahuluan penyakit virus dan vektornya pada tanaman pisang. Jakarta: Laporan Kegiatan Penelitian LPH, 1978, pp. 210-211.

295. **Summanwar, A.S. and Marathe, T.S. (1982)**: Diagnostic technique for the detection of bunchy top and infectious chlorosis in banana suckers. Curr. Sci., **51**: 47-49.

296. **Sun, S.K. (1961).** Studies on bunchy top disease of bananas. Special Publ. Coll. Agri. Taiwan, Univ., **10**: 82-109.

297. **Suresh, S and Regupathy, A. (1985).** Persistence toxicity of systemic Insecticides to the banana aphid, *Pentalonia nigronervosa* Coq. South Indian Hort., **33**: 78-82.

298. **Takada, H. (1992).** Transmission by aphid vectors of virus diseases of papaya and banana. A. Aphid parasitoids as biological control agents of vector aphids of papaya ring spot virus and banana bunchy top virus. FFTC Technical Bull., No. **132**: 1-11.

299. **Thangavelu, R. and Singh, H.P. (1996).** Status of banana bract mosaic virus and banana streak virus disease in India. (Abstr.). In: Conference on Challenges for banana production and utilization, p. 56, 24-25 Sept., 1996, NRC on Banana, Trichy.

300. **Thomas, J.E. (1991).** Virus indexing procedures for banana in Australia. In: Proc. of a Regional Technical Meeting on Diseases Affecting Banana and Plantain in Asia and the Pacific, Brisbane, Australia, 15-18 April, 1991 (Ed. Ramon V. Valmayor).

301. **Thomas, J.E., and Dietzgen, R.G. (1991).** Purification, characterization and serological detection of virus-like particles associated with banana bunchy top disease in Australia. Jour. Gen. Virol., **72**: 217-224.

302. **Thomas, J.E., Geering, A.D.W., Gambley, C.F., Kessling, A.F. and White, M. (1997).** Purification, properties and diagnoses of banana bract mosaic potyvirus and its distinction from abaca mosaic potyvirus. Phytopathology **87**: 698-705.

303. **Thomas, J.E., Geering, A.D.W., Daniells, J.W. and Mc. Michael, L.A. (1998).** Banana streak virus in Australia. In: Banana streak virus: a unique virus-*Musa* interaction. Proc. of a workshop of the PROMUSA Virology Working Group (eds. E.A. Frison and S.L. Sharrock), Montpellier, France, 19-21 January, 1998.

304. **Thomas, J.E. and Iskra—Caruana, M.L. (2000).** Diseases caused by viruses. In: D.R. Jones (ed.). Diseases of Banana, Abaca and Enset. CAB Publishing, New York. pp. 241-293.

305. **Thomas, J.E. and Magnaye, L.V. (1996).** Banana bract mosaic disease. Montpellier, France, International Network for the

Improvement of banana and Plantain (INIBAD) Musa Disease Fact Sheet No. 7, 1 p.

306. **Thompson, J.D. and Dietzgen, R.G. (1995).** Detection of DNA and RNA plant viruses by PCR and RT-PCR using a rapid virus release protocol without tissue homogenisation. J. Virol. Methods **54**: 85-95.

307. **Thottapilly, G., Dahal, G., Harper, G., Hll, R., and Lockhart, B.E.L. (1997).** Banana streak badnavirus: Development of diagnostics by ELISA and PCR. Phytopathology **87**: 597.

308. **Thottapilly, G., Dahal, G. and Lockhart, B.E.L. (1998).** Studies on a Nigerian isolate of banana streak badnavirus: I. Purification and Enzyme-linked immunosorbent assay. Annals of appl. Biol.,**132**: 253-261.

309. **Trindade, D.R., Poltronieri, L.S., Albuquerque, F.C. De, Benchimol, R.L. and Amorim, A.M. (1998).** Occurrence of cucumber mosaic virus on banana in the State of Para, Brazil. Fitopatologia Brasileira **23**: 185.

310. **Tushmereirwe, W.K., Karamura, E.B. and Karyeija, R. (1996).** Banana streak virus (BSV) and an associated filamentous virus (unidentified) disease complex of highland bananas in Uganda. Infomusa **5**: 9-12.

311. **Uppal, B.N., Varma, P.M. and Capoor, S. P. (1945).** A mosaic disease of Cardamom. Curr. Sci., **14**: 208-209.

312. **Van Hoof, H.A. (1962).** A virus disease of banana and plantain. Surin. Landb., **10**: 47-52.

313. **Vakili, N.G. (1969a).** Bunchy top disease of bananas in the central highlands of South Vietnam. Plant Dis. Reptr., **53**: 634-638.

314. **Vakili, N.G. (1969b).** Control and eradication of bunchy top disease of banana in South Vietnam. ISSAID Vietnam/Tech. Advisory Pamphlet **1**: 11 pp.

315. **Varasi, M.S. (1954).** Bunchy top of banana in Hyderabad

State. Proc. 41st Indian Science Congress, Part III, p. 256 (Abstr).

316. **Varma, P.M. and Capoor, S.P. (1958).** Mosaic disease of Cardamom and its transmission by the banana aphid (p. *nigronervosa*). Indian J. Agric. Sci., **28**: 97-107.

317. **Vasudeva, R.S. (1956).** Overseas News. Common Phytopath. News **2**: 25-29.

318. **Voelkl, W., Stechmann, D.H. and Stary, P. (1990).** Suitability of five species of Aphididae (Hymenoptera) for the biological control of the banana aphid *Pentalonia nigronervosa* Coq. (Homoptera, Aphididae) in the South Pacific. Trop. Pest Management **36**: 249-257.

319. **Vuylsteke, D.R., Chizala, C.T. and Lockhart, B.E.L. (1996).** First report of banana streak virus in Malawi. Plant Dis., **80**: 224.

320. **Vuylsteke, D.R., Hughes, J.d'A., and Rajab, K. (1998).** Banana streak badnavirus and cucumber mosaic cucumovirus in farmers fields in Zanzibar. Plant Disease **82**: 1403.

321. **Waite, B.H. (1960).** Virus diseases of bananas in Central America. Proc. Caribb. Reg. Am. Soc. Hort. Sci., **4**: 26-30.

322. **Wanitchakorn, R., Harding, R.M. and Dale, J.M. (1997).** Banana bunchy top virus DNA-3 encodes the viral coat protein. Archives of Virology **142**: 1673-1680.

323. **Wardlaw, C.W. (1934).** Banana disease 8. Notes on various diseases occurring in Trinidad. Trop. Agric. (Trinidad),**11**: 143-149.

324. **Wardlaw, C.W. (1940).** Banana diseases 13. Observations on the condition of banana plantations in the Republic of Haiti. Trop. Agric. (Trinidad), **17**: 124-127.

325. **Wardlaw, C.W. (1948).** Infectious chlorosis of banana in the Cameroons. *Nature* (Lond.) **162**: 894.

326. **Wardlaw, C.W. (1961).** Banana diseases. London: Longmans, Green and Co. Ltd., 1961.

327. **Wardlaw, C.W. (1972).** Banana Diseases, including plantain and Abaca. Longman, London, 878 pp.

328. **Wardlaw, C.W. and McGuire, L.P. (1933).** Cultivation and diseases of the banana in Brazil. Trop. Agric. Trinidad **10**: 192-197; 211-227; 255-259.

329. **Wetzel, T., Candresse, T., Macquarie, G., Ravelonandro, M. and Dunez. J (1992).** A highly sensitive immunocapture polymerase chain reaction method for plum pox potyvirus detection. J. Virological Methods **39**: 27-37.

330. **Wolcott, G.N. (1933).** An Economic Entomology of the West Indies. San Juan P.R. The Entomological Society of Puerto Rico., p. 688.

331. **Wu, M.L., Hung, T.H. and Su, H.J. (1997).** Strain differentiations of cucumber mosaic virus associated with banana mosaic disease in Taiwan. Ann. Phytopath. Soc. Japan **63**: 176-178.

332. **Wu, R.Y. (1987).** Characterization and monoclonal antibodies of the virus causing banana bunchy top. Ph.D. Dissertation, Dept. of Plant Pathology and Entomology, National Taiwan Univ., Taipei, Taiwan, ROC.

333. **Wu, R.Y. (1994).** Two circular single—stranded DNAs associated with banana bunchy top virus. Jour. Phytopath., **142**: 292- 300.

334. **Wu, R.Y. and Su, H.J. (1990a).** Purification and characterization of banana bunchy top virus. Jour. of Phytopathology **128**: 137-145.

335. **Wu, R.Y., and Su, H.J. (1990b).** Production of monoclonal antibodies against banana bunchy top virus and their use in enzyme-linked immunosorbent assay. J. Phytopath., **128**: 203-208.

336. **Wu, R.Y. and Su, H.J. (1991).** Regeneration of healthy banana plantlets from banana bunchy top virus-infected tissues cultured at high temperature. Plant Pathology **40**: 4-7.

337. **Wu, R.Y. and Su, H.J. (1992).** Detection of banana bunchy top virus in diseased and symptomless banana plants with monoclonal antibody. Tro. Agric., **69**: 397-399.

338. **Wu, R.Y., You, L.R., and Soong, T.S. (1994).** Nucleotide sequences of two circular single stranded DNAs associated with banana bunchy top virus. Phytopathology **84**: 952-958.

339. **Wu, M.L., Hung, T.H. and Su, H.J. (1997).** Strain differentiation of cucumber mosaic virus associated with banana mosaic disease in Taiwan. Ann. Phytopath. Soc. Japan **63**: 176-178.

340. **Xie, W.S. and Hu, J.S. (1995).** Molecular cloning, sequence analysis and detection of banana bunchy top virus in Hawaii. Phytopathology **85**: 339-347.

341. **Yang, L.L. (1970).** Studies on the varietal resistance to bunchy top disease of banana. J. Taiwan Agric. Res., **19**: 57-63.

342. **Yeh, H.H., Su, H.J. and Chao, Y.C. (1994).** Genome characterization and identification of viral associated ds DNA component of banana bunchy top virus. Virology **198**: 645-652.

343. **Yot-Dhauthy, D. and Bove, J.M. (1966).** Mossique du banarier. Identification et purification de diverse souches du virus. Fruits d' Outre Mer., **21**: 449-465.

344. **Zambolim, E.M., Assis, M.l.T., Zambolim, L., Venturia, J.A., and Carvalho, M.G. De (1994).** Natural infection of banana cultivar Prata (AAB) by cucumber mosaic virus in the state of Minas Gerais, Brazil. Fitopatol. Bras., **19**: 483-484.

345. **Zeck, E.H. and Eastwood, H.W. (1929).** The banana aphid (*Pentalonia nigronervosa*) (Coq.). Agric. Gaz. N.S.W., **49**: 675 p.

346. **Zhang Hai Bao, Zhu Xi Ru and Liu Hong Xian (1997).** Effect of banana bunchy top virus (BBTV) on endogenous hormones of banana plant. Acta Phytopath. Sinica **27**: 79 -83.

347. **Zhou Zhong Ju, Lin Qi Ying, Xie Lian Hui, Chen Qi Jian, Wu Zu Jian, Huang Guo Shui, Jiang Jia Fu and Zheng Guo**

Zhang (1996). Studies on banana bunchy top. IV. Control of the disease. Jour. of Fujian Agric. Univ., **25**: 44-49.

348. **Zora Singh, Jones, R.A.C. and Jones, M.G.K. (1995).** Identification of cucumber mosaic virus subgroup I isolates from banana plants affected by infectious chlorosis disease using RT-PCR. Plant Disease **79**: 713-716.

AUTHOR INDEX

SUBJECT INDEX